Mean Reversion Trading Systems

Mean Reversion Trading Systems

Practical Methods for Swing Trading

Howard B. Bandy

Blue Owl Press, Inc.

Copyright © 2013 by Howard B. Bandy. All rights reserved. No part of this publication may be reproduced, stored in a retrieval system, or transmitted in any form or by any means, electronic, mechanical, photocopying, recording, or otherwise without the prior written permission of the copyright holder, except brief quotations used in a review.

AmiBroker is a trademark of AmiBroker and Tomasz Janeczko.
Excel is a trademark of Microsoft Corporation.
Premium Data is a trademark of Norgate Investor Services.
IQ Feed is a trademark of Televent DTN.

ISBN-13: 978-0-9791838-4-3

LCCN: 2012920496

Published by
Blue Owl Press, Inc.
3700 S. Westport Avenue, #1876
Sioux Falls, SD 57106

Published 2013
Printed in the United States
19 17 16 15 14 13   10 9 8 7 6 5 4 3 2 1

Disclaimer

This book is an educational document. Nothing in this book is intended as, nor should it be construed to be, investment advice.

The views expressed herein are the personal views of Dr. Howard B. Bandy. Neither the author nor the publisher, Blue Owl Press, Inc., have any commercial interest in any of the products mentioned. All of the products described were purchased by the author at regular retail prices.

Investing and trading is risky and can result in loss of principal. Neither this book in its entirety, nor any portion thereof, nor any follow-on discussion or correspondence related to this book, is intended to be a recommendation to invest or trade mutual funds, exchange traded funds (ETFs), stocks, commodities, options, or any other financial instrument. Neither the author nor the publisher will accept any responsibility for losses which might result from applications of the ideas expressed in the book or from techniques or trading systems described in the book.

The programs used as examples have been tested and are believed to be correct. Even so, this book may contain typographical errors and other inaccuracies. Past performance, whether hypothetical, simulated, back-tested, or actual, is no guarantee of future results. Results will depend on the specific data series used. Please verify the accuracy and correctness of all programs before using them to trade.

Acknowledgements

Tomasz Janeczko, author of AmiBroker. Thank you for creating an outstanding program, and for graciously allowing use of materials published in the AmiBroker user documentation.

Dedicated to Galileo

Contents

Preface 11
    Companion Books 11
    Blog 11
    Educational Material 11
    AmiBroker Development Platform 12
    Excel Spreadsheets 12
    Intended Audience 12
    Organization of the Book 13
    The Author 13

1   Introduction 15
    Why Mean Reversion? 15
    Mean Reversion versus Trend Following ... 16
    System Overview 17
    Explanations 17
    KISS 20
    Trade-offs 21

2   Development and Analysis 23
    Technical Analysis 23
    The Signal and the Noise 24
    The Development Process 24
    Distributions 31
    The Importance of Consistency 34
    Risk 36
    Position Size 37
    Profit Potential 37
    System Health 37
    Dynamic 37

	Schematic of a Trading System	38
	Flowchart of System Development	39
	Dissonance Alert	40
	Managing Subjectivity with Objective Functions	41
	Optimization and the Curse of Dimensionality	45
	Why Traders Stop Trading	47
3	Naive System	49
	Buy After an N-Day Sequence	49
4	Development Template	67
5	Transformations	69
	Indicators	69
	The Vertical Axis	71
	The Horizontal Axis	71
	Distribution of Indicators	72
	A Standard	73
	pdfs and CDFs	73
	Transformations	75
	Normalization	76
	Re-scaling	77
	Z Transformation	80
	Out-of-range Values	81
	Softmax Function	83
	Stationarity	87
	Indicators Demystified	87
	Moving Averages	88
	Moving Average Crossover	89
	Detrended Price Oscillator	91
	Position in Range	91
	DV2	94
	Rank Transformation	94
	Diffusion Index	99
	Stochastic	104
	RSI	104
	Exotic Indicators	112
	Back to the Vertical Axis	113
	Quality Matters	113

6	Exits	117
	Logic	118
	Holding Period	124
	Profit Target	127
	Trailing Exit	133
	Maximum Loss Exit	135
	Viewing Trades as Bars	135
	Summary	138
7	Entries	141
	Market on Close of Action Bar	141
	Next Day Open	143
	Limit Order	144
	Stop Order	145
	Future Leak	145
	Anticipating Action	146
8	Controlling Risk	147
	Filters—ATR	148
	Filters—Moving Average	159
	Position Size	163
	Options	163
9	Systems	173
	3 Day High Low	174
	Regime Change	192
	Connors RSI Pullback	198
	Dual Time Frame	201
	VIX	209
	Chapter Summary	216
10	Multi-System Systems	217
	Validation	218
	Degrees of Freedom	218
	Positive Correlation	219
	Using Equity Changes for Analysis	220
	Summary	221
11	It Doesn't Work Anymore	223

References 227

Program Listings 231
    About the Programs and Listings 231
    Downloadable Code 232

Index 235

Preface

Companion Books

Mean Reversion Trading Systems is a companion to *Modeling Trading System Performance* and an extension of *Quantitative Trading Systems*. While, primarily for continuity, some of the material from each of those other books has been included in this book, the majority of this book is new material not covered in either of the others.

Quantitative Trading Systems focuses on the design, testing, and validation of trading systems. It is available at its website:
http://www.QuantitativeTradingSystems.com

Modeling Trading System Performance focuses on analysis of trading systems. Major topics include Monte Carlo simulation, estimation of risk and profit potential, techniques for dynamic determination of position size based on recent system performance. It is available at its website:
http://www.ModelingTradingSystemPerformance.com

Introduction to AmiBroker is an introduction to the AmiBroker trading system development platform. It is free for personal use, and available from its website:
http://www.IntroductionToAmiBroker.com

Blog

Dr. Bandy hosts a blog with periodic postings and articles, including links to reference material.
http://www.blueowlpress.com/WordPress/

Educational Material

This book is intended to be educational. Trading systems, portions of trading systems, and results of applying trading systems are shown to illustrate points of discussion. None of the results are from trades actually taken. All of the results are simulated. None of these systems, or the ideas behind them, are intended to be used as trading systems as they are presented. Always do your own research, system development, and validation before trading any system.

All of the ideas, methods, and techniques discussed are either original developments made by Dr. Bandy, are available from easily accessible non-copyrighted materials, or are used with permission. When known, original authors are given credit. No non-disclosure agreements have been violated.

AMIBROKER TRADING SYSTEM DEVELOPMENT PLATFORM

In order to describe trading system topics unambiguously, it is necessary that they be coded into the language of a development platform so that they can be tested. Programs in this text are coded in AmiBroker's afl language. The AmiBroker platform was chosen because, in the opinion of this author, it is the highest quality platform available to retail-level developers. Programs written in afl are clear and concise. Execution is fast. The language is capable of implementing all of the topics to be discussed.

This text is not an AmiBroker reference or instructional manual. For that, I recommend both my companion book, *Introduction to AmiBroker,* and AmiBroker's reference and tutorial materials. The AmiBroker website is:
http://www.amibroker.com/

AmiBroker Version used: 5.60

End-of-day data is provided by Premium Data.
http://www.premiumdata.net/

Intra-day data is provided by IQFeed.
http://www.dtniq.com/index.cfm

EXCEL SPREADSHEETS

Analysis of trading system results is done using Excel spreadsheets and the techniques described in *Modeling Trading System Performance.* The Excel add-in that performs the simulations is provided from MTSP's website.

INTENDED AUDIENCE

Experienced trading system developers wanting more information about methods for mean reversion and swing trading systems.

ORGANIZATION OF THE BOOK

Chapter 1 describes mean reversion systems, compares them to other types of trading systems, and gives an overview of the characteristics of the systems that are developed in this book.

Chapter 2 is a general discussion of the trading system development process, followed by a discussion of analysis of trading system results.

Chapter 3 describes a mean reversion trading system based on a simple concept and shows that mean reversion systems are feasible.

Chapter 4 provides a development template. System settings are specified, and sections for entries and exits are provided.

Chapter 5 discusses data transformations. Whether price data or an indicator, transformations are useful to change the distribution of data —converting unbounded to bounded, changing boundaries, normalizing, linearizing, etc.

Chapter 6 discusses exits. Several exits that work well for mean reversion systems are described, and suggested for general use.

Chapter 7 discusses entries. The most frequently used methods for entering trades are discussed.

Chapter 8 discusses ways to control risk. Topics include filters, position size, and options.

Chapter 9 illustrates more complete mean reversion trading systems.

Chapter 10 discusses issues related to multi-system systems.

Chapter 11 gives suggestions for dealing with systems that are no longer working as well as they once did.

THE AUTHOR

Dr. Howard Bandy:
- Has university degrees in mathematics, physics, engineering, and computer science.
- Has specialized in artificial intelligence, applied mathematics, modeling and simulation.
- Was professor of computer science and mathematics, and a university dean.
- Designed and programmed a well-known program for stock selection and timing.
- Was a senior research analyst for a CTA trading firm.

Readers who appreciate this book are encouraged to visit our websites and learn about the other materials available at
http://www.BlueOwlPress.com

1

Introduction

WHY MEAN REVERSION?

This book presents methods for the development—the design, test, validation, and analysis—of statistically sound, practical, tradable swing trading systems.

The technique used to identify the entry is often used to describe the type of trading system. Those techniques include:
- Trend following. Buy when prices have already shown an upward trend, with the expectation that they will continue to rise.
- Mean reverting. Buy when prices are unusually low, with the expectation that they will return to more normal values.
- Seasonality. Buy when historical patterns of time suggest prices will rise.
- Patterns. Buy when a pattern that often precedes a rise in price is identified.
- Cycles. Buy when a low point in a price cycle has been detected, expecting that prices will rise.

Swing trades are meant to include trades that last between one day and a few days. Each system developed in this book is intended to take either a long or a short position in a single highly liquid index or exchange traded fund (ETF). We will test the systems on a variety of tradable issues, but the intent of those tests is system validation rather than portfolio construction.

There is no question that every successful trade is a trend following trade for the period it is active. When we are long, we need a rising trend; when we are short, we need a falling trend. With swing trades, we are entering the long position when prices are recognized to be at relative lows.

MEAN REVERSION VERSUS TREND FOLLOWING

The systems described are mean reversion systems. Whether the pattern that triggers the entry is based on an indicator, a price pattern, position within a cycle, or something else, it will always be that the price has deviated significantly from the mean, and the trade is taken with the expectation that the price will revert to the mean. We will be buying low and selling higher; not buying high and selling higher, as a trend following breakout system would.

Regression to the mean is normal in many ways:
- Children of tall parents tend to be tall, but less tall than their parents.
- Children of short parents tend to be short, but less short than their parents.
- Intelligence of children, as measured by IQ, tends to be more nearly average than their parents.
- Performance in athletics, such as baseball batting averages, tends to improve after bad months and decline after good months.
- Stocks that have the highest P/E ratios in one year tend to have lower ratios the next year, and vice versa.
- Mutual funds that performed best one year tend to drop in ranking the following year, and vice versa.
- The accuracy of navy pilots landing on aircraft carriers tends to be mean reverting.

While developing systems, it is often possible to design the logic so that the resulting trades can be either mean reverting or trend following by providing switches or extending ranges of parameter values. This is done in some of the systems described. It is useful both to:
- Determine whether the best performance comes from being mean reverting or trend following.
- Automatically adjust the system when a regime change is detected.

Swing trading is active trading and has short holding periods. Some would say that, in itself, is a bad thing—that the proper technique is to buy and hold. From my point of view, each trader has the responsibility to maximize the growth of his or her account while holding risk of a drawdown to a tolerable level. Throughout this book the goal will be to develop trading systems that achieve that objective. Whether they have low or high turnover, and long or short holding periods, the results, measured in terms of profit potential for a given level of risk, will speak for themselves.

The goal is to develop systems the trader can be confident will perform well. Confidence comes from successful completion of the validation phase of the development process, and from understanding the distribution of profit and distribution of drawdown that the system is likely to provide in the future.

None of our criteria will involve purity of doctrine or adherence to traditional wisdom. The determination of success will be based on statistical evidence, applied to systems developed using sound modeling and simulation procedures.

System Overview

The target characteristics of most of the systems described in this book are:
- Trades SPY, or a similar index or ETF.
- Development period is 1/1/1999 through 1/1/2012.
- Takes only long positions.
- Trades about 24 times per year.
- Holds a few days.
- Is highly accurate with a high percentage of winning trades.
- Uses end-of-day data.
- Most entries are made at the close of trading at the closing price.
- Exits are made either at the close of a day or at an intra-day price that is pre-computed.
- Profit targets are used.
- Maximum holding periods are used.
- Trailing exits are used.
- Maximum loss exits are not used.
- System logic is generally simple.
- Data is often transformed.
- Auxiliary data series are sometimes used.

Explanations

Trades ETFs

The focus is on exchange traded funds, ETFs. The techniques described may be applicable to other tradable issues—individual equities, FOREX contracts, futures contracts, traditional mutual funds, prices of fine art, or whatever you wish—but results of testing those issues is not reported here.

We develop the systems using ETFs in large part because it is easier. Most ETFs are designed to follow an index. SPY, for example, is designed to track the S&P 500 index—500 individual stocks combined as a weighted average. The averaging process dampens the volatility of the individual issues, resulting in an index price series that is smoother than those of the individual components.

Although the system is developed using the price series of an index or ETF, the signals can be used to take positions in either that issue or in a related tradable—futures contract, one individual issue, many individual issues, another ETF, or an option.

Test period

The test period used is 1/1/1999 through 12/31/2011. Prices for many issues are roughly the same at the end of this period as they were at the beginning. This provides a 13 year period with little upward or downward bias. Systems that are profitable during periods of rising prices may be simply taking advantage of a bull market. That is not a bad thing, but it does depend on identifying the bull market. Systems that are profitable during flat periods have removed the bull market bias.

Long only

Each of the systems will be designed to identify either a long entry or a short entry, but not both. That is, each system will be either long / flat or short / flat.

There are several advantages to a system that is choosing between only two alternatives rather than three:
- The program for long / flat will be shorter, clearer, have fewer parameters, be easier to program, and run faster.
- The rules and parameters needed to decide whether to be long or flat are fewer and simpler than the rules and parameters needed to decide which of three categories—long, flat, or short.
- The accuracy of identification of long signals will be higher when there are only two categories.
- Separating long / flat systems from short / flat systems removes any temptation to treat signals as being symmetric.

Any time the number of rules and parameters is increased, there should be an increase in the value of the model. Adding logic to detect the short portion of short / flat in a long / flat / short system does nothing to increase the performance of the long trades, but does increase the likelihood of loss of generality and does increase the curve-fitness of the model.

Also read Curse of Dimensionality in Chapter 3.

Systems that are long / flat are making binary decisions—they are asking whether the correct position is long or flat. From the perspective of machine learning and pattern recognition, they are binary classifiers. They sort every pattern, no matter how it is defined, into one of two categories—long or not long.

> (Everything said about long / flat systems can also be said about short / flat systems. Discussing only one at a time makes the text of this book clearer. I am often confused by descriptions that try to include both long and short—such as "go long (short) when the indicator rises (falls) through the critical level"—so they will be avoided.)

There is a bias toward rising equity prices. Some of the contributing factors include:
- Resource extraction.
- Population growth.
- Inflation.
- Productivity improvement.
- Survivorship bias.
- Feedback from long-only investments.
- Legislative rules.

Bottoms in prices are more clearly defined than tops. It is easier to develop a system that selects long entries than one that selects short entries.

It is more common to be long than to be short. Short trades are prohibited in some accounts, such as most IRA accounts.

The logic to identify short entries is typically not just the symmetric opposite of the long entry. So isolating long and short signals produces programs that have fewer logical statements and fewer variable parameters.

Prices act differently as they are rising than as they are falling. Particularly in rising markets, price drops are often steeper and of shorter duration. While any given technique might be effective in either a long / flat or short / flat system, the specific rules and parameter values are not necessarily symmetric. If the system buys when the 3 and 9 period simple moving averages cross, the best signals to short do not necessarily come from 3 and 9 period averages, from the same method of computing the average, or even from use of a moving average crossover as the indicator.

TRADES 24 TIMES A YEAR, HOLDS A FEW DAYS

There are two seasonality patterns that occur regularly. One is related to the end of the month, the other to options expiration. In general, it is profitable to be long for a few days beginning a few days before the end of the month, and also be long about a week before options expiration Friday. These are almost reliable enough to trade as systems. But regardless of that, they do suggest that there are about 24 good opportunities to enter a long position each year. (And, perhaps, 24 opportunities to enter a short position.) Readers of my book, *Modeling Trading System Performance*, will appreciate that frequent trading is important as it allows for frequent compounding. And, short holding periods are important to limit exposure to drawdown.

You are taking in a lot of uncertain, inaccurate, or biased information, and making a certain decision—be long, flat, or short.

The goal is to trade highly liquid ETFs, such as SPY. For the most part, the data analyzed to generate the trading signals will be the daily data of those same ETFs.

Tests will be run on three groups:
- Highly liquid ETFs, including SPY, QQQ, IWM, EEM, and GLD.
- Sector ETFs, including the nine S&P sector funds XLB, XLE, XLF, XLI, XLK, XLP, XLU, XLV, and XLY.
- A group of highly liquid issues that were at approximately the same price at the beginning of the test period as they were at the end, including AEP, ALL, BAC, BK, CMCSA, COF, CSCO, DD, DELL, DIS, EWJ, F, GE, HD, HNZ, INTC, IVV, KO, MRK, MS, MSFT, PFE, QQQ, RTN, SMH, SPY, T, TWX, VZ, WY, XLU, XLV.

KISS

Some developers suggest the KISS—Keep It Simple and Straightforward—or some variation. As we will see in a later chapter, KISS can be effective. But I caution developers that simple and straightforward systems are easily discovered by others. I do recommend that systems have as few subjective decisions and as few optimizable parameters as possible. Simple does not necessarily mean common, obvious, easy, trivial, unsophisticated, ordinary, or traditional.

Rather, uniqueness, and often sophistication and elegance, is a requirement to keep the system from rapidly losing its edge as other traders begin discovering the same patterns.

I am in agreement with Nate Silver who, in *The Signal and the Noise*, recommends we should aim to be right on a high percentage of little trades rather than searching for a few big wins.

TRADE-OFFS

Mean reversion systems that trade frequently and have a high percentage of winning trades:
- Are easier to analyze to determine system health.
- Give more precise estimates of system performance.
- Have smaller gains per trade.
- Can be traded with smaller accounts.
- Can be traded at higher fraction.
- Are designed to take advantage of volatility.
- Benefit from frequent compounding.
- Through shorter holding periods, are less exposed to drawdowns.
- Are not self correcting when in losing trades.

Trend following systems that trade infrequently and have a low percentage of winning trades:
- Are much more difficult to analyze to determine system health.
- Give highly variable estimates of system performance.
- Are designed to capture larger gains from longer trends.
- Require larger trading accounts.
- Must be traded at lower fraction.
- Are self correcting when in a losing trade.
- Through longer holding periods, are exposed to higher drawdowns.

2

Development and Analysis

This chapter is an overview of the process I recommend for development and analysis of trading systems. It is independent of any particular method of trading. It is the method used to develop and analyze mean reversion systems described in this book.

TECHNICAL ANALYSIS

Technical analysis and quantitative analysis are based on the belief that several conditions are true.
1. The price and volume reflect all available and necessary information about the company, fund, or market.
2. There are patterns in the records of price and volume that regularly precede profitable opportunities.
3. We can discover those patterns.
4. Those patterns will continue to exist long enough for us to trade them profitably.
5. The markets we model are sufficiently inefficient for us to make a profit trading them.

Technical analysis began as chart analysis, and has developed a large body of subjective interpretation of chart artifacts such as flags, retracements, head-and-shoulders, and trend lines, to name just a few. We traders are all very good at selective vision—we see what we want to see. We can look at a chart and see examples of a big gain following, say, the breakout of a triangle pattern. Thinking we have found a good trade entry technique, we can define those conditions in very precise terms and have the unbiased computer search all the data for instances of that pattern. The results do indeed show a profit for the pattern we saw so clearly. Often it also shows losses from many similar patterns that we either did not see or chose not to acknowledge; and it sometimes shows signals that appear, and then disappear as additional data points are added to the chart.

QUANTITATIVE ANALYSIS

Quantitative analysis refines technical analysis by:
- Removing the judgment associated with ambiguous chart patterns.
- Defining unambiguous, mathematically precise indicators.
- Requiring that no indicator or signal may change in response to data that is received after it has been initially computed.
- Making extensive use of mathematical models, numerical methods, and computer simulations.
- Applying statistical validation techniques to the resulting trading models.

The mean reversion systems developed in this book are based on the principles of quantitative analysis.

THE SIGNAL AND THE NOISE

A trading system is defined to be a combination of a model and one or more series of data. The model contains the rules, logic, and parameters. There is at least one data series—the issue being traded. In many systems, that is both the series examined to detect patterns and the series that is traded. In some systems, auxiliary data series are used to aid in recognition of good entries and exits, or to help define conditions when trades should not be taken.

The data analyzed consists of signals and noise. The signal is that component that contains the patterns that are recognized by the logic of the model; everything not specifically treated as signal is noise, even if there are profitable trading opportunities in the noise. From the point of view of trend following with holding periods of months or longer, swing trades are within what it considers to be the noise. Similarly, day traders and scalpers find profitable trades within what swing traders consider to be the noise. This book focuses on swing trades.

THE DEVELOPMENT PROCESS

All traders need confidence that their system will be profitable when traded. For any trader, the way to build confidence is to practice.

One advantage that traders who use mechanical systems have over those who use graphical methods is that the mechanical system is (or, at least, can be) objective, rule-based, judgement free, and testable.

You will be writing a trading system, testing it using historical data, then following the rules to place actual trades. The period of time, and the data associated with that time, that you use to develop and test your system is called the in-sample period and the in-sample data. The period of time, and the data associated with that time, that follows the in-sample period and has never been tested or evaluated by the system is called the out-of-sample data. Actual trades are always out-of-sample.

The process of validating a trading system is one of observing the profitability and behavior of the system in the out-of-sample period after it leaves the in-sample period.

The transition you make going from testing your system in-sample to trading it out-of-sample is one data point in the validation of your system. Your confidence level will be much higher if you have observed many of these transitions. The walk forward process automates these practice steps.

Risk of drawdown, profit potential, and position size are all determined by analysis of the out-of-sample trade results.

The outline presented here is designed to build your confidence that your system will be profitable. For a more extensive discussion of trading system development, including expansion of the topics presented in this outline, please see *Quantitative Trading Systems* and *Modeling Trading System Performance*.

1. OBJECTIVE FUNCTION

An objective function is a metric of your own choosing that you use to rank the relative performance of two or more alternative trading systems. The phrase "of your own choosing" is critical. Your objective function must accurately reflect what is important to you. If you prefer long holding periods and infrequent trading, your objective function must rate systems that hold for weeks to months higher than systems that hold for a few days. If you prefer high equity gain without regard to drawdown over lower, but smoother, gain your objective function must reflect that.

I believe the psychology of trading experts who try to help traders become comfortable with the systems they trade have it backwards. If you decide ahead of time what you want, and design trading systems that satisfy your wishes, and if you have the confidence built through the validation process, you are guaranteed to be comfortable with your system.

Trading system development platforms, including AmiBroker, report the score for many metrics with each test run. If one of these standard, built-in metrics meets your needs, use it as your objective function. If you want something else, you can create a custom metric and use it. For many people, rewarding equity growth while penalizing drawdown is important. If you agree, consider using one of the standard metrics that does that:

- RAR/MDD (risk adjusted annual rate of return / maximum drawdown).
- CAR/MDD (compound annual rate of return / maximum drawdown).
- K-ratio.
- Ulcer Performance Index.
- RRR (risk-reward ratio).

2. What to trade

This book focuses on highly liquid indexes and ETFs, such as SPY, QQQ, and IWM. Much of the analysis is done using daily data. Some systems compute signals at the close of trading for execution at the close of that bar, some for execution the next day, either at the open or at a pre-computed intra-day price.

3. Design the system

A trading system is a combination of a model and one or more sets of data. The model is contained in the code you write. The data is the price data of the ticker symbol your code processes. The model contains the intelligence. It is looking for the patterns in the data and testing the profitability of buying and selling.

A model consists of several parts: filter or setup, entry, one or more exits, position size, portfolio composition, and so forth. In much of the literature, the entry is emphasized. But the other components are equally important.

A model is a static representation of a dynamic process. Once you are done coding and testing, the model does not change. It may be cleverly designed and have self-adapting parameters, but it is still static. The market being modeled is dynamic and ever changing. Your model is looking for a particular pattern or set of conditions, after which it expects a profitable trade. As long as the model and the market remain in synchronization, the system will be profitable. When the two fall out of sync, the system will be less profitable or unprofitable.

It is useful to think of the data as comprised of two components—signal and noise. The pattern your model is looking for is the signal portion of the data. Everything your model does not recognize is the noise portion of the data. Your goal, as a designer of trading systems, is to accurately recognize the signal and ignore the noise.

Our hope is that:
- We can build a model,
- That recognizes some inefficiency,
- And use that model to trade profitably,
- As long as the model and reality stay in sync.

4. IN-SAMPLE PERIOD

There are two views about how much time and data should be used to develop the system.

Some want a long time. They feel that the system will experience a variety of conditions and be better able to handle changes in the future. The risk is that conditions vary so much during the in-sample period that the system will not learn any of it well.

Some want a short time. They feel that the system will be better able to synchronize itself and learn very well. The risk is that the system may learn a temporary pattern that does not persist beyond the in-sample period. Some of that closer fit is to the noise. A system that fits the noise and performs well in-sample, but does not perform well out-of-sample, is described as a system that is curve-fit, or over-fit, to the data.

The proper length of the in-sample period is impossible to state in general. It is very much a function of both the model and the data. My view is that the length of the in-sample period should be as short as is practical. The only way to determine the length of the in-sample period is to run some tests.

5. OUT-OF-SAMPLE PERIOD

The length of the out-of-sample period is: As long as the model and the market remain in sync and the system remains profitable. There is no general relationship between the length of the out-of-sample period and the length of the in-sample period.

6. OPTIMIZE

To optimize means to search a lot of alternatives and choose the best of them. Optimization is not a bad thing; in fact, it is a necessary step in system development. If you were to think up a new trading system and write the code for it, it would include logic and parameters that were first guesses. You could test that system as it stands, and then either trade it or erase it and start over. But you are unlikely to do that. More likely, you will try other logic and other parameters values. If you are trying a few hand-picked alternatives, you might as well run an optimization and try thousands of alternatives. Give yourself a chance to find the best system.

Anything you would consider changing in your system—a different parameter value or an alternative logic statement—is a candidate to be optimized. The value of optimization is not so much in searching thousands of alternatives—it is in ranking them and choosing the best. The intelligence is in the objective function.

7. WALK FORWARD RUNS

While there is nothing special about optimizing, there is something very special about a walk forward run.

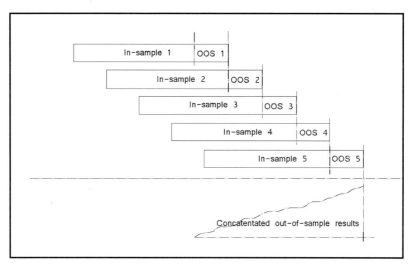

Figure 2.1 -- Walk Forward Part 1

The walk forward process is several iterations of:
1. Optimize in-sample.
2. Choose the best according to your objective function.

3. Use those values and test out-of-sample.
4. Step forward by the length of the out-of-sample period.

Continue this process until you have used the last full in-sample period.

8. BEST ESTIMATE SET OF TRADES

There are two sets of trades produced by the walk forward process—those from all of the in-sample periods, and those from all of the out-of-sample periods. The out-of-sample trades are the reference set. They are the *best estimate* of future performance of the system.

9. EVALUATE OUT-OF-SAMPLE RESULTS

Analyze the best estimate set of trades. Either send the system on to analysis and potentially trade the system or send it back to development.

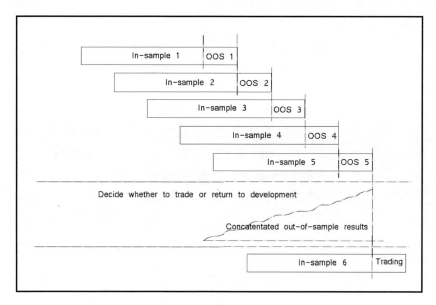

Figure 2.2 -- Walk Forward Part 2

Note the importance of having an objective function you trust. The set of parameters used to test out-of-sample are the parameters that are at the top of the results list, after the list has been sorted by your objective function. The process is automatic and objective. All the decisions were made in earlier steps. You will never even see what the second choice is; the first choice is always used.

Each walk forward step is one data point you will use in the validation of your system.

10. DETERMINE POSITION SIZE

Proper position size is required to meet two goals:
- Maximize account growth.
- Hold drawdown to an acceptable level.

Using the best estimate set of trades, along with your personal and subjective risk tolerance, and the Monte Carlo simulation techniques described in *Modeling Trading System Performance*, determine the position size.

Please note an important point. The system processes the data by applying the rules to generate the entry and exit signals—the system produces the trades. But the system does not determine position size. Position size is a function of trading performance, of the synchronization between the data and the logic, and of the system health. Position size is determined independently of the trading system.

11. TRADE THE SYSTEM

Using the set of parameters that are at the top of the list after the last optimization, buy and sell when the system gives signals.

Take all the signals. If you have some way to decide which signals to take and which to skip, that logic belongs in the trading system and should go through the validation process.

On the last day of what would have been the out-of-sample period, re-optimize. Pick the top-ranked parameter values and continue to trade.

12. MONITOR THE SYSTEM HEALTH

Each trade signaled after development has finished, and each trade you actually take, is an out-of-sample trade. Compare the statistics for your trades with statistics for the out-of-sample results from the walk forward runs.

If your results drop below what is statistically expected, reduce your position size, possibly stop trading the system. Either paper-trade it to see if it recovers; or re-optimize ahead of schedule, then paper-trade it and observe its performance.

Refer to *Modeling Trading System Performance* for detailed discussion of determining whether the system is working or broken, and for techniques for computing the best position size in order to maximize account growth while holding drawdown to a level within the tolerance of the trader.

13. Resynchronize

As trades are taken, modify the best estimate set. Either add the new trades, or replace the oldest trades with the newest trades. As the distribution of those trades changes, the correct position size changes. When the performance deteriorates, as detected by either statistical tests or by the recommended position size becoming smaller, resynchronize the model to the data by repeating the validation steps.

Summary

There are no guarantees. The best we can hope for is a high level of understanding and confidence gained through the validation process.

Distributions

Our goal in system development is to combine a set of logic with a data stream in such a way that profitable trading opportunities can be identified. Short of knowledge of the future, the best estimates of future performance are obtained when two conditions are both true.
- Tests of the system are run using data that has not been used in the development of the system. That is, out-of-sample data and out-of-sample results.
- The data used is representative of the data that will be analyzed and traded.

Emanuel Derman, in his book *Models.Behaving.Badly*, states that models are simplifications; and simplifications can be dangerous. System developers should avoid simplification of data representation. Whenever possible, use distributions rather than a limited number of scalar values.

The information content that describes a trading system over a given period of time can be described in many ways. The following list is in decreasing order of information.
- Reality. Trades, in sequence, that actually result from applying the system.
- List of trades, in time sequence.
- List of trades.
- Distribution of trades.
- Four moments describing the distribution.
- Mean and standard deviation.
- Mean.
- Direction.

Probability and statistics distinguish between population and sample. The population is all items of the type being analyzed; the sample is a subset of the population that has been observed. The purpose of developing trading systems is to learn as much as possible about the population of trades that will occur in the future and make estimates of future performance. The results of testing trading systems form the sample that will be used to make those estimates.

Reality

Reality cannot be known in advance. Estimating reality—the population—is the purpose of system validation. Reality is the logic of the system processing the future data series.

List of trades, in time sequence

The list of trades, in time sequence, that results from processing a set of that data that is similar to the future data is the best estimate we can obtain of reality. There is one of these sequences for each unique set of test data and each set of logic and parameter values. Using these results to estimate future profitability and risk depends on the degree of similarity between the test data and the future data.

List of trades

The list of trades, ignoring time sequence, relaxes the assumption of the trades occurring in a particular sequence. It provides a set of data with, hopefully, the same characteristics as the future data, such as amount won or lost per trade, holding period, intra-trade drawdown, and frequency of trading. Selecting trades from this list in random order gives opportunity to evaluate the effects of similar conditions, but in different time sequence.

Distributions

A distribution can be formed using any of the metrics of the individual trades. The distribution is a further simplification since there are fewer (or at most the same number of) categories for the distribution than for the data used to form it. For example, a distribution of percentage gain per trade is formed by sorting the individual trades according to gain per trade, establishing ranges and bins, assigning each trade to a bin, and counting the number of trades in each bin. A plot of the count per bin versus gain per bin gives a plot of the probability mass function (often called the probability density function, pdf).

Four moments

Distributions can be described by their moments. The four moments most commonly used are mean, variance, skewness, and kurtosis. Depending on the distribution, some of the moments may be undefined.
- Mean. The first moment. The arithmetic average of the data points.
- Variance. Second moment. A measure of the deviation of data points from the mean. Standard deviation is the positive square root of variance.
- Skewness. Third moment. A measure of the lopsidedness of the distribution.
- Kurtosis. Fourth moment. A measure of the peakedness and tail weight of the distribution.

Mean and standard deviation

Mean and standard deviation are commonly computed and used to describe trade results. They can be used in the definition of metrics such Bollinger Bands, z-score, Sharpe ratio, mean-variance portfolio, etc.

Mean

The mean gives the average of the values. Mean can be computed in several ways, such as arithmetic mean and geometric mean. The mean is the single-valued metric most often associated with a set of data.

Direction

Direction of a trade describes whether it was a winning trade or a losing trade. Direction is meant to represent any way of describing the trades in a binary fashion. Other ways might be whether the result was large or small in absolute value, or whether the maximum favorable excursion met some criterion, etc.

Stay high on the list

With each step down this list, a larger number of data points are consolidated into a smaller number of categories, and information is irretrievably lost. Knowing only the information available at one level makes it impossible to know anything definite about the population that could be determined at a higher level. Working with only the mean tells us nothing about variability. Working with only mean and standard deviation tells us nothing about the heaviness of the tails. Using the four values of the first four moments enables us to calculate some information about the shape of the population, but nothing about the lumpiness or gaps that may exist.

APPLICATION TO MODELING TRADING SYSTEMS

The Monte Carlo simulation technique repeatedly builds and analyzes a sequence of trades. Each sequence represents a fixed time horizon, say two years. The number of trades in that sequence is the number of trades that can be expected to occur in two years. Each trade is selected from either a trade list or a distribution that represents a trade list. When enough trades to cover two years have been selected, the equity curve, win-to-loss ratio, maximum drawdown, and other metrics of interest can be calculated–just as if this sequence was a two-year trade history.

After generating and analyzing many, usually thousands, of two-year sequences, form and analyze the distributions of the results. Two of the most important metrics of results are account growth and maximum closed trade drawdown. Others that might be of interest include win-to-loss ratio, maximum intra-trade drawdown, etc.

The important points are:
- Use the best data available. Out-of-sample data that best represents the anticipated future data that will be traded.
- Use the highest representation of the data possible. That is, either a list of trades or a detailed distribution of trades.
- Using simulation techniques, generate many possible trade sequences.
- Analyze the distributions of the results of the simulation.
- Calculate and use metrics based on the distributions to judge the usefulness of the system.

FURTHER READING

There are several books listed on the Resources page of my website that have excellent discussion of the importance of distributions and the dangers of using only single values. If you can read just one, begin with Sam Savage, *The Flaw of Averages: Why We Underestimate Risk in the Face of Uncertainty*.

THE IMPORTANCE OF CONSISTENCY

Backtest results depend on the fit between the model and the data. As developers, our first goal is backtests that are highly profitable with smooth equity curves. We hope these results are indicative that the logic is accurately identifying profitable patterns—distinguishing the signal from the noise. But until the system has been validated we cannot tell

Development and Analysis

what portions of the results are simply data mining.

The validation process, particularly the walk forward process, provides us with a set of trades that are the best available estimate of how the system would perform as it is periodically resynchronized and traded. Even at that, some of the out-of-sample results were detected as patterns, but are random values that just happen to fit the rules; there is no profitable trade following them. Further, the walk forward process provides a single sequence of trades. If the future data is identical to the past data, that sequence might be repeated. If the future is similar (has the same distribution of trades), but trades occur in a different order, the equity curve will have a different shape.

Figure 2.3 shows ten equity curves, each created by randomly selecting 50 trades (the average number of trades in a two year period) from the best estimate set associated with the naive system described in Chapter 3. Each trade is made with all available funds. The average return per trade is 0.41%. The average equity after 50 trades is expected to be $122,701 ($100000*1.0041^{50}$). The average of the ten on the chart is the dark dashed line, with a terminal equity of $127,650. If you were trading the system for two years, your results could be curve A, or B, or C—all with equal probability.

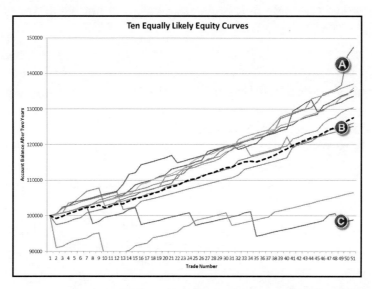

Figure 2.3 - Ten equally likely equity curves

The consistency of the trades determines the predictability of the account equity. That, in turn, determines risk, position size, and profit potential.

As market conditions change, and trades taken by the system are added to the best estimate set, the distribution of future equity curves changes accordingly. Those actual trades are important for monitoring system health and dynamically adjusting position size.

RISK

Risk, as I define it here, is measured as percentage drawdown from the highest balance of the trading account. Risk, and profit potential, begins with the individual trade, then progresses to the sequence of trades.

Each trader must understand the characteristics of the system being traded. As related to risk, intra-trade adverse excursion is important.

Your system has exit rules that will eventually generate a signal to close every trade. If you are always willing to hold positions until the system gives the signal, then the drawdown is based on the profit or loss of closed trades. If you would override your system and close a trade because it had a large intra-trade loss, then the drawdown is based on intra-trade profit or loss.

I understand that there will be unusual and unexpected conditions. But I recommend that as many of those as possible be anticipated, code to generate the appropriate signals be added to the system logic, and the effect tested. The code that implements what would be an override of the logic is a maximum loss exit.

The extend to which the intra-trade drawdown is important depends on the goal of the system. A swing trading system expects to hold a short period of time, is often taking positions that are oversold, and regularly has an intra-trade loss at some point during the trade. A trend following system expects to hold a long period of time. While it is waiting for the large profit trade, it is exposed to larger intra-trade drawdowns. Swing traders are less likely to override their systems due to an intra-trade drawdown, and, consequently, closed trade results are appropriate for analyzing risk.

As I discuss later in this chapter, the primary reason traders stop trading a system is a drawdown in account equity that exceeds their tolerance.

Risk tolerance is personal and subjective. It is expressed as a combination of two numbers.

One is the maximum drawdown the trader is willing to accept before taking the system offline. This might be 10, 15, or 20%. If a trading account had grown to a balance of $200,000, a drawdown of 10% is a drawdown of $20,000. Drawdowns always increase over time. If a system is likely to have a 10% drawdown over a one year horizon, over a two year horizon under the same circumstances it will have about a 14% drawdown. Drawdown increases in proportion to the square root of holding period for individual trades, and in proportion to the square root of the forecast period for system performance estimates.

The other is the confidence the trader wants about the drawdown limit. That is expressed in terms of the percentile of the distribution of drawdowns. A desire to be 90% confident means that if the system trades for ten two year periods, the maximum drawdown will not exceed 10% in nine of those periods.

Every trading system has some probability of experiencing a sequence of losing trades that exceed the trader's tolerance. The best estimate set of trades, together with the Monte Carlo simulation described in *Modeling Trading System Performance*, can be used to quantify that probability.

POSITION SIZE

Once the risk is quantified, the maximum position size—the maximum fraction of the account used for each trade—that can be used without the drawdown exceeding a tolerance limit can be computed. The position size is independent of the trading system itself. It can be computed from the size of the trading account, the best estimate set of trades, and the trader's risk tolerance.

POTENTIAL PROFIT

Given the maximum position size, the distribution of account equity can be computed and translated back to compound annual rate of return, CAR. The metric of CAR / Maximum Drawdown is useful in deciding whether a system is worth trading.

SYSTEM HEALTH

System performance and system health fluctuates as the data changes and the degree of synchronization between the model and the data changes. Account growth and drawdown change accordingly.

DYNAMIC

Every system is dynamic. As trades are made, those trades should be added to the best estimate set, and the updated set used to reassess risk and adjust position size.

Schematic of a Trading System

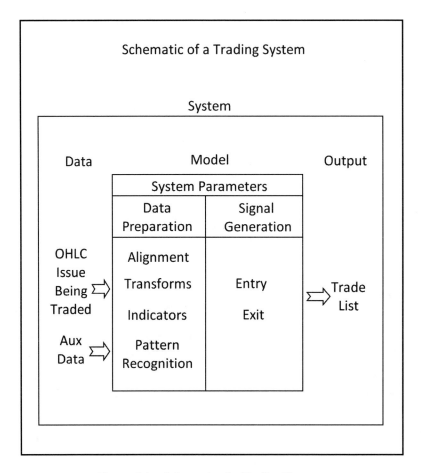

Figure 2.4 -- Schematic of a Trading System

Flowchart of System Development and Trading Management

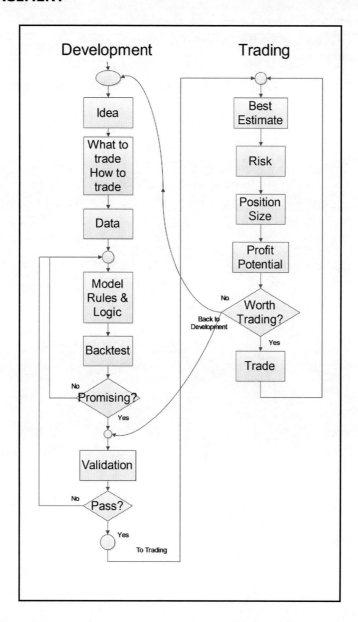

Figure 2.5 -- Flowchart of System Development and Trading Management

DISSONANCE ALERT

Expert advice says both:
- Ignore the in-sample results!
- Have confidence in the in-sample results!

How can both be reasonable?

During the early phases of development, the logic and parameters are adjusted to give the best fit to the in-sample data. If that fit is to the signal portion of the data, and the characteristics of the data remain consistent in the following out-of-sample data, then the out-of-sample performance will be good.

The only way to fit the model to the data is through in-sample optimization. The only way to verify that the fit is profitable is by testing out-of-sample data.

In the final analysis, the system relies on good in-sample results to select the rules, determine the parameters, and identify the signals. But you cannot have confidence that the system is reliable without good out-of-sample results.

MANAGING SUBJECTIVITY WITH OBJECTIVE FUNCTIONS

Every day, traders must make decisions:
- What to trade.
- When to enter.
- How large a position to take.
- When to exit.
- Whether the system is healthy.

Discretionary traders acknowledge the subjectivity of their decisions and draw on experience. Systematic traders use objective functions designed to identify important decision criteria and quantify them.

Objective functions are important in two phases of trading systems design:
- To rank alternative systems during system development.
- To decide the size of the next position during system analysis.

RANKING ALTERNATIVE SYSTEMS

A trading system is a set of computations, logic statements, and parameter values that comprise a set a rules that identify profitable trading patterns and give buy and sell signals.

There are an infinite number of possible systems. In order to make the process manageable, relatively simple systems are designed to focus on specific trading ideas, such as trend following, mean reverting, seasonality, etc.

For any one of these ideas, there are many alternatives. A trend following system might have logic that looks at breakouts, the crossing of two moving averages, or the projection of a regression. For each of these there are numeric parameters such as the lengths of the moving averages, or magnitude of breakout. There might be multiple rules to exit a position, such as logic, trailing exit, profit target, and / or maximum loss stop.

Designing a trading system is an iterative process of:
- Modify the logic and parameters.
- Test the performance.

Each set of logic and parameters creates a new trading system – one of the alternatives to be evaluated. The developer needs to decide which is *best*.

Best is subjective. The purpose of the objective function is to provide an objective metric that represents the subjectivity of the developer's definition of best.

Depending on the preference of the trading system developer, determination of best includes evaluation of profitability, risk, trading frequency, holding period, execution efficiency, percentage of winning trades, and any other measurable feature.

A decade ago, the reporting from test runs made by trading system development platforms was limited to basic metrics such as net profit and maximum drawdown. Today, many platforms provide both a wide range of sophisticated metrics and the capability for the developer to design custom metrics. This gives the developer the opportunity to incorporate subjective judgments into a custom objective function–a metric computed using scores and weights for each of the evaluation criteria.

To insure that the objective function accurately reflects the preferences of the developers, it must be calibrated. The calibration process is:

1. Pick a time period and tradable issue and apply a trading system, generating a set of trade results and equity curve.
2. Perform Step 1 several times, using different logic and different parameter values. Using the same time period and the same issue, generate the reports and equity curve for each alternative.
3. Print out the reports and equity curves, one to a page, and lay them out on a table or floor.
4. Arrange them into order according to your subjective preference—from the one you would be most comfortable trading to the least. If necessary, make up and include hypothetical results to fill in gaps in performance and to represent results typical for your trading.
5. Analyze the results, paying particular attention to the features that are important to you. Create a list of these features. Create a measurement scale for each of the features, and assign relative weights to each feature. Set up the measurement scale so higher values are preferred. The sum of all the measurement scores and weights is your objective function.
6. Return to the trade reports and calculate the objective function score for each report. This step should be very easy. If it is not, return to Step 5.
7. If the order the reports were placed in using your subjective judgment is the same as the order of the objective function, you have created an objective function that reflects your subjective preferences. You are done.
8. If the two orderings are different, modify your objective function

by adding or removing terms and modifying weights. Return to Steps 5, 6, and 7 until you are satisfied.

When you are satisfied, program your objective function and use it to objectively rank alternatives. Whenever you find that you prefer some alternative other than the one ranked highest, return to the design of your objective function and refine it. The goal is to have a high degree of confidence that trading results that have high objective function scores are results you like, and the systems that produced them are systems you would be confident trading.

I favor functions that reward account growth, reward smoothness, and penalize drawdown. Several well documented functions that include these criteria are:
- RAR/MDD (risk adjusted annual rate of return / maximum drawdown).
- CAR/MDD (compound annual rate of return / maximum drawdown).
- K-ratio.
- Ulcer Performance Index.
- RRR (risk-reward ratio).

Upon completion of calibration, the designer can be confident that the trading system ranked highest among all the alternatives is the one he prefers. This is useful during the design and testing phases. It is critically important during the walk forward testing in the validation phase because the model with the highest objective function score is automatically used to trade the out-of-sample period.

For more discussion about objective functions, see *Quantitative Trading Systems*.

DECIDING THE SIZE OF THE NEXT POSITION

The trading system that results from the design, testing, and validation provides a single set of trades with single mean, single standard deviation, single terminal wealth, single maximum drawdown.

These results will be repeated as the system is traded only if the future price series is exactly the same as the historical series used during development. In order to estimate profit potential and risk it is important to consider the distribution of potential results.

Modeling future performance, including evaluating system health, estimating profit potential and estimating risk is based on:

1. Using the set of trade results that, in the judgment of the developer, best represents the trades that are likely to occur in the future.
2. Using Monte Carlo simulation techniques to create equally likely trade sequences.
3. Analyzing the distributions of profit and drawdown resulting from the trade sequences.
4. Comparing both the magnitude and probability of both the profit potential and drawdown with the trader's personal tolerance for risk and desire for profit to determine system health and position size.

An objective function formed by taking the ratio of the Compound Annual Rate of Return (CAR) at some confidence level to the Maximum Drawdown (MaxDD) at some confidence level is very useful in deciding whether a system is worth trading.

This process, including the software necessary to run the Monte Carlo simulations, is described in detail in *Modeling Trading System Performance.*

Position size is intimately related to system health and changes dynamically as the synchronization between the logic of the trading system and the price series it is processing changes. System health must be monitored during trading, and position size revised as necessary.

Optimization and the Curse of Dimensionality

On a forum I regularly follow, Yahoo AmiBroker Forum, a message complained that the AmiBroker trading system development platform was unable to handle an optimization that had 10 parameters, each with 20 values. Whether that poster was serious about the numbers 10 and 20 or not, the question is worth considering.

A large number of parameters, say 10, each with a large number of evaluation points, say 20, leads directly to the "curse of dimensionality." Assuming that it takes one nanosecond to evaluate each alternative, and that everything works as it should and the run does finish (no power failure, memory failure, cpu failure, hard disk failure, flood, earthquake, premature death, etc), exhaustive evaluation of the $n = 10^{20}$ points will have taken about 3100 years. At 50 lines per 12 inch page, the listing of the results will be 65 light years long. (The nearby universe is described as being those bodies within about 15 light years.)

Each result gives the value of an objective function. Sorting the results into ascending or descending order according to the value of the objective function puts the set of values that are "best" at the top of the list. Assuming a sorting algorithm that takes Order $(n*\ln(n))$ is used, it will take an additional 46 times 3100 years to sort them.

Now make an individual run using each of the sets of values near the top to gain some confidence that the alternative ranked first is the one that is preferred. Maybe an hour or two more?

Just in time to give a signal for tomorrow's trade?

What is a reasonable number of alternatives to evaluate?

I recommend beginning by giving a considerable amount of thought and experimentation to design and testing of the objective function. Have confidence that the alternative with the highest objective function score is the one that is preferred. Or at least that it is acceptable—not letting the search for perfect get in the way of finding a satisfactory solution.

Run some tests using your system, your objective function, your data, on your computer. How long does it take to generate and sort 1000 or 10,000 alternatives? How many of these test runs do you expect to make before deciding on the logic and parameters that will be used? How many hours are you willing to allocate to the process? Working the arithmetic backwards will give an estimate of the number of alternatives that can realistically be tested. Continuing to work backward, you can compute the size of the parameter space that can be exhaustively processed in that amount of time. Make some trial runs using both exhaustive and non-exhaustive methods. Compare the results that rank best from each set of runs. Decide whether you can accept the results of the non-exhaustive method. If so, use it; if not, modify the afl code so that the number of parameters and number of evaluation points for each create an examination space that can be processed in the time you have allocated.

WHY TRADERS STOP TRADING

Assume a trader has a method – mechanical, discretionary, or a combination of both – that she has been using successfully. Also assume that she understands both herself and the business of trading, and wants to continue trading. Why would she stop trading that system?

Here are a few possibilities:
1. The results are too good.
2. The results are not worth the effort.
3. The results are not worth the stress.
4. She has enough money.
5. There is a serious drawdown.

1. Results are too good, and she is afraid that this cannot possibly continue. Her system – any system – works when the logic and the data are synchronized. There are many reasons why systems fail and should be taken offline, but a sequence of winning trades should be seen as a success.
   She should continue trading it until one of the other reasons to stop happens.
2. The results are not worth the effort. There is not much gain, but not much loss either. On balance, the time, energy, and resources would be more productively applied doing something else.
   The problem is with the system. She can return the system to development and try to refine it; or take it offline and periodically review its performance.
3. The results are not worth the stress. Performance is satisfactory, but at a high cost – worry and loss of sleep. Regardless of the position size indicated by the distribution of risk, the positions being taken are too large.
   She should either reduce position size or have someone else execute the trades.
4. She has enough money.
   No matter how good a system is, there is always a risk of serious loss. When she has reached her goal, she should retire.
5. There is a serious drawdown. The magnitude of the drawdown needed for it to be classified as serious is subjective. Among my colleagues and clients, those who manage other people's money typically want drawdown limited to single digits. Those trading

their own money may be willing to suffer drawdowns of 15 or 20 percent.

But for every person there is a level of drawdown at which he stops trading the system – preferably while the account still has a positive balance.

My view is that experiencing a large drawdown is the primary reason people stop trading a system.

What causes a large drawdown and how should the trader react to it?
1. The system is broken.
2. There was an unexpected sequence of losing trades.
3. The system is out of synch.
4. The position size is too high.

As the account balance drops from an equity high into a drawdown, it is not possible to determine which is The reason.

All of the reasons are true to some extent. A system that is broken breaks because the logic and the data become unsynchronized, causing an unexpected sequence of losing trades and at a time when position size was too high for conditions.

The solution is two-fold:
1. Continually monitor system performance and system health.
2. Continually modify position size to reflect recent performance.

During the trading system development process, a baseline of system performance is established. The out-of-sample trades from the walk forward phase is a good source of this data. Personal risk tolerance and system risk, taken together, determine position size for that system performance. As system performance changes, position size must also change.

Position size varies in response to system health.

Conclusion

Do not continue to trade a system that has entered a serious drawdown expecting that it will recover. It may recover on its own; it may require readjustment; or it may be permanently broken and never work again.

Take it offline and either observe it until recent paper-trade results demonstrate that it is healthy again, or send it back to development.

The correct position size for a system that is broken is zero.

3

Naive System

Before expending a lot of energy designing and testing mean reversion systems, we will examine a naive system to see whether SPY, and other highly liquid ETFs and equities exhibit mean reversion or trend following behavior.

As readers of my companion book, Modeling Trading System Performance, will recognize, I recommend systems that trade frequently, hold for a short period of time, and have a high percentage of winning trades. The systems described in this chapter are not finished systems. They are intended to be exploratory and should not be traded without extensive testing and validation by the trader.

BUY AFTER AN N-DAY SEQUENCE

Defining a day as either a rising day or a falling day based on whether the most recent close is higher or lower than the previous day's close, we can test the feasibility of buying after a sequence of rising days or a sequence of falling days.

Every trading system involves making some subjective decisions. For this system, they include:
- Issue traded — SPY
- Date range — 1/1/1999 to 1/1/2012
- Indicator — A sequence of rising or falling daily prices.
- Action — At the close of the signal bar.
- Commission and slippage — none.
- Positions — Long only.
- Objective function — CAR / MDD.

- Initial equity — $100,000.
- Maximum open positions — 1
- Position size — Fixed at $10,000

Every system has some parameters. For this system they include:
- N — The number of consecutive closes in the same direction.
- Direction — Rising (1) or falling (0).
- HoldDays — Maximum holding period in days.
- ProfitTarget — Exit intra-day when this percent profit can be realized.

The rules are:
- Buy — N consecutive closes in the same direction.
- Sell — At the close of the maximum holding period, or at the profit target, whichever is reached first.

Listing 3.1 shows the AmiBroker code.

```
//   BuyAfterAnNDaySequence.afl
//

SetOption( "ExtraColumnsLocation", 1 );
SetOption ( "CommissionMode", 2 );  // $ per trade
SetOption( "CommissionAmount", 0 );
SetOption( "InitialEquity", 100000 );
SetPositionSize( 10000, spsValue );
MaxPos = 1;
SetOption( "MaxOpenPositions", MaxPos );
SetTradeDelays( 0, 0, 0, 0 );
BuyPrice = Close;
SellPrice = Close;

//   Define a day as rising based on the closing price
Rising = C > Ref( C, -1 );
Falling = C < Ref( C, -1 );

//   The number of days in the sequence
N = Optimize( "N", 3, 1, 7, 1 );

//   Direction.  1 == Rising, 0 == Falling
Direction = Optimize( "Direction", 1, 0, 1, 1 );

//   Exit variables
//   Maximum holding period
HoldDays = Optimize( "HoldDays", 4, 1, 7, 1 );
//   Profit target
ProfitTarget = Optimize( "ProfitTarget", 0.4, 0.2, 4, 0.2 );

//   Detect an N day sequence

if ( Direction == 1 )
{
    NDaySequence = Sum( Rising, N ) == N;
}
```

Naive System

```
    else
    {
        NDaySequence = Sum( Falling, N ) == N;
    }

    Buy = NDaySequence;
    Sell = 0;

    ApplyStop( stopTypeProfit, stopModePercent, ProfitTarget );
    ApplyStop( stopTypeNBar, stopModeBars, HoldDays );

///////////////   end   ///////////////
```

Listing 3.1 -- Buy After an N-Day Sequence

The optimization statements call for testing of:
- N days from 1 to 7.
- Direction both Rising and Falling.
- Holding Period from 1 day to 7. In AmiBroker, both the entry day and the exit day are counted. Since entry is made at the close of one day and the earliest exit is at a profit target the following day, the shortest period reported will be for a 2 day holding period.
- ProfitTarget from 0.2% to 4.0%.

Figure 3.1 shows the result of the optimization, sorted by CAR / MDD (Compound Annual Return / Maximum Drawdown).

N	Direction	HoldDays	ProfitTar...	CAR	Net Profit	CAR/...	K-Ratio	# Trades	Sharpe Ratio	Max. Sys % Dr...	Profit Fact...	Avg Profit/L...	% of Winn...	Avg % Profit/Lo...
3	0	7	0.4	0.71	9,640.55	0.97	0.0993	207	5.61	-0.74	5.45	46.57	95.17	0.47
3	0	4	0.4	0.60	8,104.05	0.93	0.1244	208	3.88	-0.65	3.61	38.96	92.79	0.39
4	0	3	1.4	0.41	5,492.78	0.90	0.0471	82	3.72	-0.46	2.75	66.99	76.83	0.67
3	0	3	0.4	0.57	7,653.24	0.90	0.1006	209	3.92	-0.63	3.20	36.62	88.04	0.37
3	0	4	0.2	0.57	7,691.04	0.88	0.1110	215	4.05	-0.65	4.90	35.77	94.88	0.36
3	0	5	0.4	0.60	8,037.38	0.88	0.1001	208	3.84	-0.68	3.41	38.64	93.27	0.39
3	0	7	0.2	0.65	8,714.72	0.87	0.0914	214	5.35	-0.74	6.92	40.72	96.73	0.41
3	0	5	0.2	0.56	7,515.45	0.86	0.1111	215	4.05	-0.65	4.37	34.96	94.88	0.35

Figure 3.1 -- Optimization results

The parameters are in the first four columns in the same order they occur in the program. Namely, N, Direction, HoldDays, ProfitTarget.

Note that the best result has parameter values of 3, 0, 7, 0.4. Buy at the close of a day that completes a sequence of 3 falling days, take a profit of 0.4% as soon as possible, exit at the close of the 7th day (6th day after entry). And the second best, which has a better K-ratio, exits on day 4.

Figure 3.2 shows the equity curves and drawdown curves for the best system (with the seven day hold).

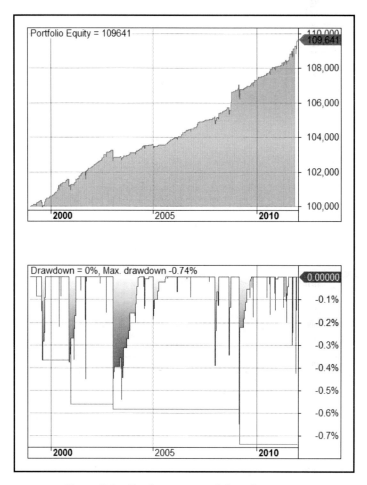

Figure 3.2 -- Equity curve and drawdown curve

Figure 3.3 shows the statistics for the best system.

	All trades	Long trades	Short trades
Initial capital	100000.00	100000.00	100000.00
Ending capital	109640.55	109640.55	100000.00
Net Profit	9640.55	9640.55	0.00
Net Profit %	9.64 %	9.64 %	0.00 %
Exposure %	1.00 %	1.00 %	0.00 %
Net Risk Adjusted Return %	961.66 %	961.66 %	N/A
Annual Return %	0.71 %	0.71 %	0.00 %
Risk Adjusted Return %	70.90 %	70.90 %	N/A
All trades	207	207 (100.00 %)	0 (0.00 %)
Avg. Profit/Loss	46.57	46.57	N/A
Avg. Profit/Loss %	0.47 %	0.47 %	N/A
Avg. Bars Held	2.65	2.65	N/A
Winners	197 (95.17 %)	197 (95.17 %)	0 (0.00 %)
Total Profit	11804.58	11804.58	0.00
Avg. Profit	59.92	59.92	N/A
Avg. Profit %	0.60 %	0.60 %	N/A
Avg. Bars Held	2.38	2.38	N/A
Max. Consecutive	40	40	0
Largest win	606.78	606.78	0.00
# bars in largest win	2	2	0
Losers	10 (4.83 %)	10 (4.83 %)	0 (0.00 %)
Total Loss	-2164.03	-2164.03	0.00
Avg. Loss	-216.40	-216.40	N/A
Avg. Loss %	-2.16 %	-2.16 %	N/A
Avg. Bars Held	8.00	8.00	N/A
Max. Consecutive	1	1	0
Largest loss	-461.06	-461.06	0.00
# bars in largest loss	8	8	0

Figure 3.3 -- Statistics

Discussion

The code includes a variable that represents direction. If a long position is taken as the price is rising, that is trend following. If a long position is taken as the price is falling, that is mean reverting. So the program represents two different trading systems and gives us the opportunity to test and compare trend following (when Direction is 1) and mean reversion (when Direction is 0) in the same run.

The default optimizer is exhaustive, so all possible combinations were tested. (AmiBroker also supports non-exhaustive optimizations / searches, and these will be used in later examples to reduce execution time.)

980 combinations of parameter values were tested for rising prices, and 980 for falling.

All of the best alternatives shown in Figure 3.1 have a Direction of 0, meaning take a long position after a sequence of falling prices — a mean reversion system.

Buying when prices have been rising is trend following. 164 of the trend following alternatives showed a net profit, 816 showed a net loss. The best trend following system had parameters of 1, 1, 7, 0.6. It entered a long position after a single rising day, took profit at 0.6%, and exited on day 7.

Figure 3.4 shows the equity chart and drawdown chart for the best trend following system. Compare with Figure 3.2, which shows the same information for the best mean reversion system.

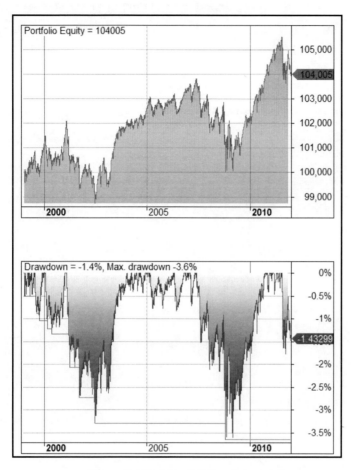

Figure 3.4 Best trend following system

Buying when prices have been falling is mean reverting. 912 of the mean reverting alternatives showed a net profit, 68 were not profitable.

VALIDATION

Having 93% of the alternatives tested profitable shows that the system is fairly robust. But these results come from testing all available data. No data was reserved for out-of-sample testing, which is generally a poor practice and over estimates profit and under estimates risk. In later studies better validation techniques, including walk forward testing, will be used.

These tests suggest that SPY has been mean reverting over the recent past. Expanding the date range to include the entire history of SPY (in a study not shown here), it has been mean reverting since it was originally published in 1993 and continues to be in summer 2012.

We can test the system using other tradable issues. There is no guarantee that every system will work for every issue, nor is there a requirement that it must. But it is interesting and valuable to know which issues tend to be mean reverting and which tend to be trend following.

Two watchlists have been prepared and will be used for cross-tradable testing throughout this book.

One is a list of 14 highly liquid ETFs:

SPY	QQQ	IWM	EEM
GLD	XLB	XLE	XLF
XLI	XLK	XLP	XLU
XLV	XLY		

The other is a list of 32 issues that meet two criteria:
- They are highly liquid. They are among the 100 most liquid issues as of 2012.
- Their price as of 1/1/2012 is within a few percent of their price 1/1/1999. Their net buy and hold performance for 13 years is roughly flat — there is no bias to being long or short over that period.

They are a mixture of ETFs and common stocks. Their ticker symbols are:

AEP	DELL	IVV	SMH
ALL	DIS	KO	SPY
BAC	EWJ	MRK	T
BA	F	MS	TWX
CMCSA	GE	MSFT	VZ
COF	HD	PFE	WY
CSCO	HNZ	QQQ	XLU
DD	INTC	RTN	XLV

The 3, 0, 7, 0.4 values found when studying SPY were used to test these two groups. Figure 3.5 shows the results of trading each issue alone, sorted by CAR/MDD. 36 of the 42 issues were profitable. None of the 6 unprofitable issues had serious losses, and even those 6 had winning trade percentages in the high 80 percent range.

Ticker	CAR... %	Net Profit	Net % Profit	Max. Sys % Drawdown	K-Ratio	# Trades	Avg Profit/Loss	Avg % Profit/Loss	% of Winners	Avg Bars Held
SPY	0.97	9,640.55	9.64	-0.74	0.0993	207	46.57	0.47	95.17	2.65
MSFT	0.81	10,232.41	10.23	-0.92	0.0771	230	44.49	0.44	93.91	2.64
QQQ	0.53	11,254.72	11.25	-1.56	0.0426	203	55.44	0.55	95.57	2.49
CMCSA	0.49	11,065.57	11.07	-1.66	0.0806	242	45.73	0.46	96.28	2.46
MS	0.38	16,094.66	16.09	-3.06	0.0392	260	61.90	0.62	92.69	2.72
PFE	0.38	7,471.77	7.47	-1.46	0.0474	224	33.36	0.33	91.07	2.88
EWJ	0.28	6,466.12	6.47	-1.74	0.0411	179	36.12	0.36	87.15	3.22
XLV	0.23	5,149.15	5.15	-1.66	0.0525	190	27.10	0.27	89.47	3.07
XLK	0.23	6,422.80	6.42	-2.11	0.0321	201	31.95	0.32	92.04	2.78
VZ	0.22	4,806.60	4.81	-1.63	0.0437	206	23.33	0.23	89.32	2.90
INTC	0.21	7,448.68	7.45	-2.59	0.0186	231	32.25	0.32	92.64	2.65
MRK	0.21	3,529.75	3.53	-1.26	0.0369	217	16.27	0.16	91.24	2.90
XLF	0.20	7,574.72	7.57	-2.83	0.0717	206	36.77	0.37	94.17	2.79
GE	0.20	5,830.88	5.83	-2.22	0.0444	223	26.15	0.26	92.83	2.81
IVV	0.18	4,194.40	4.19	-1.78	0.0414	164	25.58	0.26	91.46	3.12
XLY	0.18	6,821.97	6.82	-2.83	0.0299	214	31.88	0.32	91.12	3.08
HNZ	0.17	3,186.34	3.19	-1.39	0.0532	196	16.26	0.16	88.78	3.14
T	0.14	5,587.63	5.59	-2.94	0.0252	211	26.48	0.26	91.00	2.99
GLD	0.13	2,886.50	2.89	-1.72	0.0177	88	32.80	0.33	90.91	3.09
EEM	0.12	2,915.36	2.92	-1.88	0.0252	108	26.99	0.27	89.81	3.15
BK	0.12	4,996.51	5.00	-3.17	0.0192	221	22.61	0.23	91.86	2.90
IWM	0.12	4,252.58	4.25	-2.79	0.0262	177	24.03	0.24	92.66	2.86
HD	0.12	4,611.44	4.61	-2.79	0.0417	232	19.88	0.20	94.40	2.67
XLP	0.11	2,626.53	2.63	-1.80	0.0292	148	17.75	0.18	87.16	3.19
F	0.08	5,670.83	5.67	-5.31	0.0254	230	24.66	0.25	91.74	2.73
SMH	0.07	3,326.79	3.33	-3.72	0.0207	219	15.19	0.15	93.15	2.67
COF	0.07	3,243.73	3.24	-3.70	0.0152	230	14.10	0.14	91.74	2.77
XLU	0.06	2,006.59	2.01	-2.52	0.0233	181	11.09	0.11	90.06	2.92
DELL	0.06	2,284.32	2.28	-3.01	0.0144	208	10.98	0.11	90.38	2.90
DD	0.06	2,434.56	2.43	-3.08	0.0130	228	10.68	0.11	90.79	2.76
XLE	0.06	2,129.36	2.13	-2.69	0.0350	187	11.39	0.11	89.30	2.87
ALL	0.05	2,159.10	2.16	-3.30	0.0081	202	10.69	0.11	90.10	2.98
KO	0.04	1,767.71	1.77	-3.58	-0.0007	197	8.97	0.09	87.82	3.27
CSCO	0.03	2,570.11	2.57	-6.06	0.0108	201	12.79	0.13	92.54	2.77
DIS	0.02	1,164.32	1.16	-4.66	0.0065	212	5.49	0.05	91.51	2.89
XLI	0.00	5.94	0.01	-2.52	0.0074	178	0.03	0.00	86.52	3.17
AEP	-0.02	-1,542.10	-1.54	-7.17	-0.0017	207	-7.45	-0.07	89.86	2.91
TWX	-0.02	-1,367.36	-1.37	-5.53	-0.0233	253	-5.40	-0.05	87.75	3.08
WY	-0.02	-1,405.86	-1.41	-6.73	-0.0218	226	-6.22	-0.06	90.27	2.88
RTN	-0.03	-4,401.83	-4.40	-10.37	0.0060	205	-21.47	-0.21	89.27	3.06
BAC	-0.03	-3,661.21	-3.66	-11.05	-0.0189	221	-16.57	-0.17	88.69	2.87
XLB	-0.05	-2,929.34	-2.93	-4.94	-0.0117	194	-15.10	-0.15	86.60	3.11

Figure 3.5 -- Naive mean reversion results

Comparison with ideal trend following

Returning to analysis of SPY, it is interesting to compare the naive mean reversion system to some ideal trend following systems.

The zigzag indicator uses a single numerical argument — the minimum percentage change between peaks and valleys. Applied to closing prices, a zigzag of a certain percentage, say 5%, forms a series of trends where every upward or downward segment is at least 5% and has no retracements greater than 5%. If a system had perfect knowledge of future prices, it could identify the zigzag bottoms as they were being formed. Assume such a system took a long position on the close of the first day following a bottom, then sold using the same holding period and profit target rules as the naive mean reversion system. The percentage was adjusted until the number of trades and percentage exposure of the ideal zigzag system roughly matched that of the naive mean reversion systems. A zigzag percentage of 1.80% was selected. Figure 3.6 shows the equity curve and drawdown curve for the result.

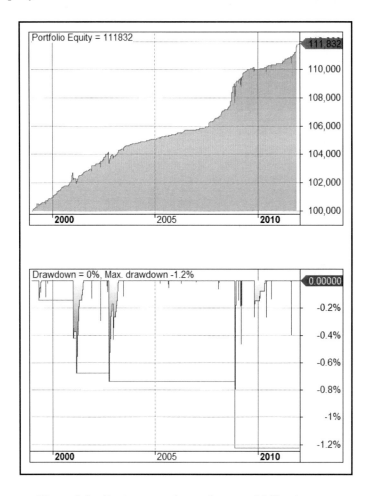

Figure 3.6 -- Equity curve for perfect trend following system

Figure 3.7 shows the summary statistics.

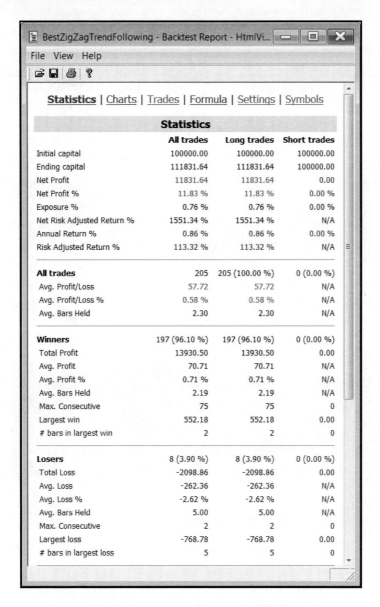

Figure 3.7 -- Statistics for perfect trend following system

EXTENSION OF MEAN REVERSION SYSTEM

Continuing analysis of the mean reversion system based on a sequence of days with falling prices, Figure 3.8 shows a chart of SPY for a six month period in 2011 with upward arrows below the bars that complete a sequence of three falling days. The system described takes a single position at the first occurrence of three falling days. There are times when a long position has not been closed, but additional falling days would cause additional signals. See Figure 3.8 and the multiple arrows in the circled bars.

Figure 3.8 -- Signals for entries using 3 falling days

Figure 3.9 shows the trade results, based on the sequence of three falling days, for the period in Figure 3.8.

Symbol	Trade	Date	Price	Ex. date	Ex. Price	% chg	Profit	% Profit	Shares	Position ...	Cum. Profit	# bars	Profit/b...	MAE	MFE
SPY	Long (profit)	1/10/2011	126.98	1/11/2011	127.488	0.40%	40.00	0.40%	78.7526	10,000.00	8,102.90	2	20.00	-0.61%	0.40%
SPY	Long (profit)	2/24/2011	130.93	2/25/2011	131.48	0.42%	42.01	0.42%	76.3767	10,000.00	8,144.91	2	21.00	-0.94%	0.42%
SPY	Long (profit)	3/16/2011	126.18	3/17/2011	128	1.44%	144.24	1.44%	79.2519	10,000.00	8,289.14	2	72.12	-0.71%	1.89%
SPY	Long (n-bar)	4/11/2011	132.46	4/20/2011	133.1	0.48%	48.32	0.48%	75.4945	10,000.00	8,337.46	8	6.04	-2.23%	0.75%
SPY	Long (profit)	5/4/2011	134.83	5/6/2011	135.369	0.40%	40.00	0.40%	74.1675	10,000.00	8,377.46	3	13.33	-1.34%	0.67%
SPY	Long (profit)	5/17/2011	133.17	5/18/2011	133.703	0.40%	40.00	0.40%	75.092	10,000.00	8,417.46	2	20.00	-0.79%	0.40%
SPY	Long (profit)	5/24/2011	131.95	5/25/2011	132.478	0.40%	40.00	0.40%	75.7863	10,000.00	8,457.46	2	20.00	-0.19%	0.59%
SPY	Long (n-bar)	6/3/2011	130.42	6/14/2011	129.32	-0.84%	-84.34	-0.84%	76.6754	10,000.00	8,373.12	8	-10.54	-2.58%	0.77%

Figure 3.9 -- Trade list for signals using 3 falling days in Figure 3.8

The program was modified so that entries required a sequence of 4, 5, and 6 falling days, and a backtest was performed for each. Figures 3.10, 3.11, and 3.12 show the trades for those systems, respectively, for the period covered in Figure 3.8.

Symbol	Trade	Date	Price	Ex. date	Ex. Price	% chg	Profit	% Profit	Shares	Position ...	Cum. Profit	# bars	Profit/b...	MAE	MFE
SPY	Long (profit)	4/12/2011	131.47	4/13/2011	132.08	0.46%	46.40	0.46%	76.063	10,000.00	2,961.57	2	23.20	-0.37%	0.46%
SPY	Long (profit)	5/5/2011	133.61	5/6/2011	134.94	1.00%	99.54	1.00%	74.8447	10,000.00	3,061.11	2	49.77	-0.44%	1.00%
SPY	Long (profit)	6/6/2011	129.04	6/7/2011	129.7	0.51%	51.15	0.51%	77.4954	10,000.00	3,112.26	2	25.57	-0.13%	1.02%
SPY	Long (profit)	6/8/2011	128.42	6/9/2011	128.934	0.40%	40.00	0.40%	77.8695	10,000.00	3,152.26	2	20.00	-0.19%	0.60%

Figure 3.10 -- Trade list for 4 falling days

Symbol	Trade	Date	Price	Ex. date	Ex. Price	% chg	Profit	% Profit	Shares	Position ...	Cum. Profit	# bars	Profit/b...	MAE	MFE
SPY	Long (profit)	4/13/2011	131.46	4/15/2011	131.986	0.40%	40.00	0.40%	76.0688	10,000.00	2,461.41	3	13.33	-0.91%	0.55%
SPY	Long (profit)	6/7/2011	128.96	6/9/2011	129.476	0.40%	40.00	0.40%	77.5434	10,000.00	2,501.41	3	13.33	-0.60%	0.86%

Figure 3.11 -- Trade list for 5 falling days

Naive System

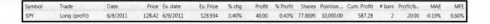

Symbol	Trade	Date	Price	Ex. date	Ex. Price	% chg	Profit	% Profit	Shares	Position...	Cum. Profit	# bars	Profit/b..	MAE	MFE
SPY	Long (profit)	6/8/2011	128.42	6/9/2011	128.934	0.40%	40.00	0.40%	77.8695	10,000.00	587.28	2	20.00	-0.19%	0.60%

Figure 3.12 -- Trade list for 6 falling days

These are seven additional trades that were not signaled by the original code. All seven were profitable.

Note that the trade entered on June 3 using the original 3 day sequence resulted in a loss of 0.84%. If longer sequences are recognized and multiple positions allowed, several additional trades would have been signaled. The four trades entered on June 6 and 8 from a 4 day sequence, June 7 using a 5 day sequence, and June 8 using a 6 day sequence resulted in a total gain of 1.71%, recovering from the 0.84% loss.

Figure 3.13 shows three equity and drawdown curves. They represent using 4, 5, and 6 day sequences of falling closing prices to enter a long trade, respectively.

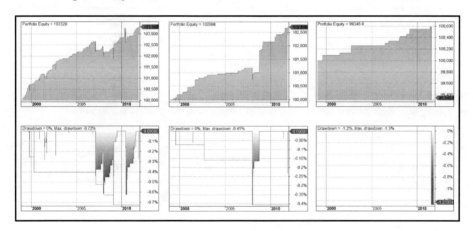

Figure 3.13 -- Equity curves for 4, 5, 6 day sequence

In my opinion, these should be treated as separate systems, not as "scaling-in." Each of the systems should be evaluated on its own merit.

The original code was modified to allow multiple positions and to enter on any sequence of three or more falling days. Figure 3.14 shows the resulting equity and drawdown curves.

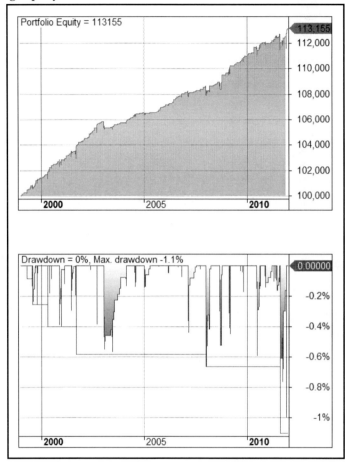

Figure 3.14 -- Equity curve allowing multiple positions

Naive System

Figure 3.15 shows the associated statistics.

	All trades	Long trades	Short trades
Initial capital	100000.00	100000.00	100000.00
Ending capital	113154.90	113154.90	100000.00
Net Profit	13154.90	13154.90	0.00
Net Profit %	13.15 %	13.15 %	0.00 %
Exposure %	1.52 %	1.52 %	0.00 %
Net Risk Adjusted Return %	868.18 %	868.18 %	N/A
Annual Return %	0.96 %	0.96 %	0.00 %
Risk Adjusted Return %	63.07 %	63.07 %	N/A
All trades	319	319 (100.00 %)	0 (0.00 %)
Avg. Profit/Loss	41.24	41.24	N/A
Avg. Profit/Loss %	0.41 %	0.41 %	N/A
Avg. Bars Held	2.66	2.66	N/A
Winners	305 (95.61 %)	305 (95.61 %)	0 (0.00 %)
Total Profit	18625.85	18625.85	0.00
Avg. Profit	61.07	61.07	N/A
Avg. Profit %	0.61 %	0.61 %	N/A
Avg. Bars Held	2.37	2.37	N/A
Max. Consecutive	46	46	0
Largest win	606.78	606.78	0.00
# bars in largest win	2	2	0
Losers	14 (4.39 %)	14 (4.39 %)	0 (0.00 %)
Total Loss	-5470.95	-5470.95	0.00
Avg. Loss	-390.78	-390.78	N/A
Avg. Loss %	-3.91 %	-3.91 %	N/A
Avg. Bars Held	9.00	9.00	N/A
Max. Consecutive	2	2	0
Largest loss	-889.11	-889.11	0.00
# bars in largest loss	9	9	0

Figure 3.15 -- Statistics allowing multiple positions

Listing 3.2 shows the AmiBroker code.

```
//    BuyAfterAnNDaySequenceMultiPosition.afl
//
```

```
SetOption( "ExtraColumnsLocation", 1 );
SetOption ( "CommissionMode", 2 ); // $ per trade
SetOption( "CommissionAmount", 0 );
SetOption( "InitialEquity", 100000 );
SetPositionSize( 10000, spsValue );
MaxPos = 7;
SetOption( "MaxOpenPositions", MaxPos );
SetBacktestMode( backtestRegularRawMulti );
SetTradeDelays( 0, 0, 0, 0 );
BuyPrice = Close;
SellPrice = Close;
//   ObFn == K-ratio, CAR/MDD, expectancy

//   Define a day as rising based on the closing price
Rising = C > Ref( C, -1 );
Falling = C < Ref( C, -1 );

// The number of days in the sequence
N = Optimize( "N", 3, 1, 7, 1 );

//   Direction.  1 == Rising, 0 == Falling
Direction = 0; // Optimize( "Direction", 0, 0, 1, 1 );

//   Exit variables
//   Maximum holding period
HoldDays = Optimize( "HoldDays", 4, 1, 7, 1 );
//   Profit target
ProfitTarget = Optimize( "ProfitTarget", 0.4, 0.2, 4, 0.2 );

//   Detect an N day sequence

if ( Direction == 1 )
{
    NDaySequence = Sum( Rising, N ) >= N;
}
else
{
    NDaySequence = Sum( Falling, N ) >= N;
}

Buy = NDaySequence;

Sell = 0;

ApplyStop( stopTypeProfit, stopModePercent, ProfitTarget );
ApplyStop( stopTypeNBar, stopModeBars, HoldDays );

//   Plots
Plot( C, "C", colorBlack, styleCandle );
shapes = IIf( Buy, shapeUpArrow, shapeNone );
shapecolors = IIf( Buy, colorGreen, colorWhite );
PlotShapes( shapes, shapecolors );

////////////////   end   ////////////////
```

Listing 3.2 -- N-Day Multiposition System

EXTREME OVERSOLD

Look in particular at the large loss by the final trade in the 6 day sequence in Figure 3.13. While there are only 12 trades in the 13 year period signaled by a sequence of 6 falling days, too few to draw conclusions, it is often the case that the most extreme oversold conditions are poor entries. One colleague gave the analogy of taking a cold remedy -- one teaspoon helps, two teaspoons helps a lot, three teaspoons makes him worse.

SUMMARY

We have discovered a simplistic, naive mean reversion system that could be traded. Its results are realistic and achievable. It is applicable to a wide range of tradable issues. Its results are roughly comparable to a nearly ideal trend following system that requires perfect knowledge of the future, giving us confidence that mean reversion systems are practical alternatives to trend following systems.

The remainder of this book expands development of mean reversion systems.

4

Development Template

For the most part, equity curves and performance statistics shown in this book are the results of backtests made over the period 1/1/1999 through 1/1/2012, using SPY as the tradable, taking only long positions. Many of the systems were tested much more extensively, including over different time periods and different tradable issues. Some of the systems are very robust and passed walk forward validation.

As I explain in much more detail in my book, *Modeling Trading System Performance*, it is important to use a set of trade results that are, to the best of the system developer's ability and belief, the best estimate of the future performance of the system. In order to avoid distortion of trade results, each trade made is the same size. Compounding, and computing the growth of the trading account, takes place in the analysis phase, and is done outside the testing environment.

I often begin development of a new system with a small template that has the code to set the system parameters. Listing 4.1 is an example of the template I used for much of the development reported in this book.

```
//   Template MOC.afl
//
//   Template for mean reversion trading systems.
//   Signals generated at the close
//   for action at the close.
//

//   System settings

SetOption( "ExtraColumnsLocation", 1 );

OptimizerSetEngine( "cmae" );

SetBacktestMode( backtestRegular );
SetBacktestMode( backtestRegularRawMulti );

SetOption( "initialequity", 100000 );
MaxPos = 1;
SetOption( "maxopenpositions", MaxPos );
SetPositionSize( 10000, spsValue );
SetOption ( "CommissionMode", 2 );  // $ per trade
SetOption( "CommissionAmount", 0 );

SetTradeDelays( 0, 0, 0, 0 );
BuyPrice = SellPrice = Close;
//ShortPrice = CoverPrice = Close;
//   ObFn == K-ratio, CAR/MDD, expectancy

//   User Functions

//   Global variables and parameters

//   Indicators

//   Buy and Sell rules

//   Plots

Plot( C, "C", colorBlack, styleCandle );

//   Explore

///////////////////////////   end  ////////////////
```

Listing 4.1 - Program Template

As you read the programs, you will see how having the sections of the program organized can be helpful.

5

Transformations

INDICATORS

Entry and exit rules can be based on a variety of ideas, limited only by the creativity of the system designer. For example:
- Patterns in prices.
- Day-of-the-month seasonality.
- Values of indicators.

This chapter discusses use of indicators—numerical values obtained by mathematical calculations based on price or volume—and their use to define rules.

For any rule to be effective, there must be a consistent and quantifiable relationship between the rule and the trade related to it. Each of the systems developed in this book takes either a long position or a short position. The quality of the rule associated with the decision to become or remain flat, or to enter either a long position or a short position, is determined by the profitability and risk of that position—by the changes in price following the rule.

Beginning with raw price and volume data, trading systems transform and interpret that data in a series of steps with the purpose of increasing the balance of our trading account.
1. Raw data
2. Transformed data / indicator
3. Trading rules
4. Trading signals
5. Trades
6. Account balance

Design of good indicators begins with some research and perfect knowledge of the future.

PLOT THE DATA

Transformations are used so it is easier to get the answer we want. For example, to identify extremes and turning points more clearly. We want clarity.

Some of the best data analysis begins with a plot of the data and visual inspection. Eventually, formula-based transformations will be needed. But plotting, charting, graphing, and visual inspection between steps will help understanding of the data and selection of the best tools.

DEFINE THE GOAL

Before polishing the magic lantern, we should decide what we want the Genie to tell us. A graph of the perfect relationship between an indicator and its trade would look like the one in Figure 5.1. The horizontal axis shows indicator values; the vertical axis shows subsequent price change.

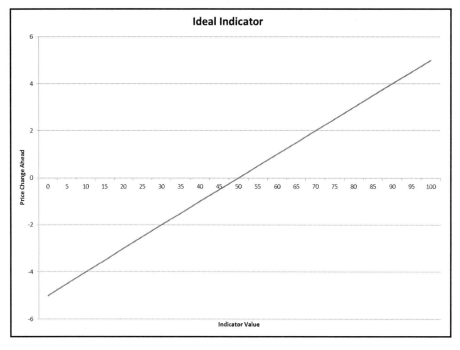

Figure 5.1 -- Ideal indicator

Low indicator values correspond to price declines, high indicator values correspond to price increases, and the amount of price change is proportional to the value of the indicator.

The vertical axis

As indicators are related to trades, the vertical axis (the "y" axis) measures the value of the trade. It could be a price change, or it could be some other metric. Higher values are better.

The vertical axis of Figure 5.1 is labeled "Price Change Ahead." For a profitable long trade, we want that to be a positive number. But it is important to define both the trading system and the way we will be executing the trades.

Traders with day jobs may only be able to process data in the evening for execution the next day. Fund managers may find it necessary to make all trades at the close of the regular trading session. Part-time traders may desire to use end-of-day data, to compute prices at which actions will be signaled, then use intra-day limit and stop orders for execution. Full-time traders, able to monitor the market throughout the trading day, may want to use intra-day data. Traders confident in their system and in their system-to-broker interface may want to use automated systems. Deciding what to predict may be more complex than the close-to-close change one day ahead.

If the system uses daily data, is strictly mechanical, enters at the close, exits at the close, and ignores intra-trade drawdowns, then the price change between entry and exit is the difference between closing prices, and the vertical axis is labeled correctly.

If entry or exit could be at a price other than the close, deciding what to predict is more complex. We will return to the vertical axis and other trade metrics in a few pages.

The horizontal axis

The horizontal axis is labeled "Indicator Value," and the range shown is 0 to 100. A range of values is required, but there are many considerations.

The indicators used in trading systems are artificial. When we choose a traditional indicator, or design a new one, we hope it has some predictive power—that is can either:
- Identify conditions that have a low probability of continuing, suggesting a mean reversion trade will be profitable.
- Identify conditions that have a high probability of continuing, suggesting a trend following trade will be profitable.

The value of the indicator determines the category of the signal—to be long or to be flat. We are not concerned with whether the indicator fits the Normal distribution, the Uniform distribution, the Binary distribution, or any other well defined and well known distribution. What we desire most from an indicator is clarity, accuracy, consistency, and ultimately profitability.

We can, and will, study the relationship between the value of the indicator and the change in price over the next few bars. If it helps, and it often does, we will modify the indicator—transform the indicator—to improve that relationship. Before we begin, we will learn a little more about indicators and transformations.

Distribution of Indicators

Some analysts differentiate indicators based on their use, categorizing some as useful in measuring the strength of price trends, others as useful in measuring the extent of price overextension. This book does not make that distinction. In my opinion, each indicator is evaluated in terms of its ability to identify profitable trades at tolerable risk.

Bounded

By their nature, some indicators are bounded—their lower and upper values cannot exceed values fixed by the definition of the indicator. The position-in-range indicators, such as stochastic oscillator, Williams %R, and percent rank are examples.

Unbounded

Other indicators are unbounded—their upper and lower values have no predetermined limits. Z-score, Bollinger %B, and CCI are examples.

A feature of unbounded indicators is that they can form peaks and valleys that extend beyond the recent range of values. Analysts who study divergences make use of the peaks and valleys of unbounded indicators.

Some of the transformations described in the following pages can convert unbounded indicators to bounded, and bounded to unbounded.

Distributions and Range of values

There are three general shapes of distributions of indicators:
- Uniform-like. Some indicators, such as rank and percent rank, have a distribution of values that is always uniform—distributed evenly throughout the range.
- Normal-like. Others, such as z-score, have distributions that are

always non-uniform, are concentrated in the mid-range and have few data points at the extremes.
- Bimodal. Still others, such as RSI, have distributions that depend on the data and / or the parameters of the indicator. When short lookback periods are used, RSI values are separated into two clusters of values at the two extremes.

Among all indicators, the ranges of indicator values vary widely in the maximum and minimum values. For some all values are positive, others all are negative, still others can be either positive or negative. For some a high value represents an overbought condition, for others it is an oversold one.

You may only care about one end of the distribution—the one that generates signals for the long positions the specific system takes, or the other end that generates signals for short positions a different system takes.

A Standard

There are far too many variations in the format of indicators. In some cases we will use an indicator in its traditional form. But in most cases—particularly when combining or comparing indicators—it would be useful to have indicators conform to a standard format. No single standard works well in all applications, so some compromise is necessary. Many of the indicators in common use are bounded with a minimum value of 0 and a maximum value of 100. RSI and percent rank are two examples.

> *When we need to standardize an indicator, we will transform it so that it is bounded, and has a range of 0 to 100. Lower values will be associated with oversold conditions (price is below the mean) and higher values with overbought conditions (price is above the mean).*

PDFs and CDFs

Recall that distributions have two representations—probability density function (pdf) and cumulative distribution function (CDF). The graph commonly used to illustrate the Normal distribution, shown in Figure 5.2, is a pdf—a probability density function.

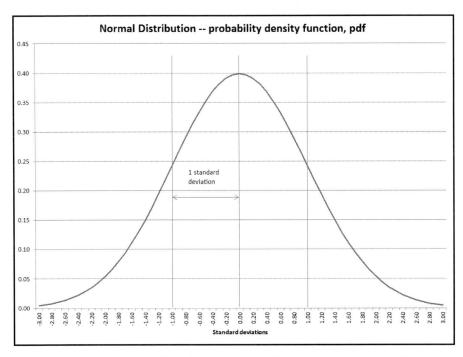

Figure 5.2 - Normal Distribution - probability density function (pdf)

The horizontal axis shows the range of values over which the function is defined. The height of the graph at any point represents the relative magnitude of the number of data points that occur at that point or in that range. For a Normal distribution, the mean (arithmetic average), median (middle value of a sorted list of values), and mode (most frequent value) all coincide at the peak in the middle. Theoretically, the line describing the pdf extends to the left and right indefinitely, never dropping completely to zero, and never touching the axis. There is, however, a practical limit. For a *standard* Normal distribution—one with a mean of 0.0 and a standard deviation of 1.0—most (99.7%) of the data points lie within three standard deviations of the mean—between -3.0 and +3.0.

The formula that defines the Normal pdf for mean μ and standard deviation σ is:

$$f(x, \mu, \sigma^2) = \frac{1}{\sigma\sqrt{2\pi}} e^{-\frac{1}{2}\left(\frac{x-\mu}{\sigma}\right)^2}$$

There is a cumulative distribution function, CDF, associated with every probability density function, pdf. The CDF for a given point represents, as its name suggests, the cumulative percentage from the left extreme to that point. In calculus terms, the CDF is the integral of the pdf, and the pdf is the derivative of the CDF. In many circumstances, the distribution is more easily examined and the transformation more clearly visualized when the CDF form is used. Figure 5.3 shows the CDF of the Normal distribution corresponding to the pdf in Figure 5.2.

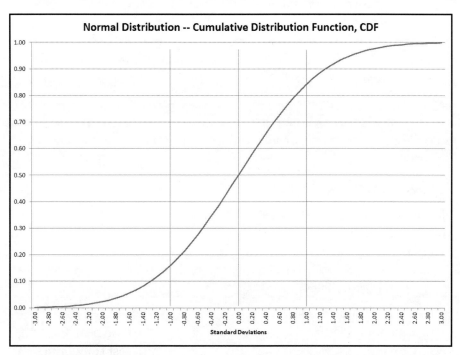

Figure 5.3 - Normal Distribution -- Cumulative Distribution Function (CDF)

Transformations

A transformation is a mathematical operation that takes the value for each data point in an input set of data and computes a corresponding value for the output set of data.

When we talk about transforming an indicator, we are talking about changing the shape of the pdf or, equivalently, the CDF. The transformed distribution has a higher or lower percentage of values in some areas, typically at the extremes, than the original indicator.

There are an infinite number of transformations. Think of transforming an indicator, or any data series, as measuring that indicator using a different scale on the ruler; or as focusing a camera before taking the picture. A common everyday transformation is converting temperature

from degrees Fahrenheit to degrees Celsius. John Tukey, an American mathematician who made many contributions to data analysis in the late 20th century, called them *re-expressions*. Examples in our context:
- Input might be price and output the RSI(price). Indicators are themselves transformations.
- Input might be the RSI value, output the rank of the RSI value. The transformation changes the distribution of the indicator to make the signals clearer.
- Input might be OHLC price bars, output a metric of trade profit and risk. The transformation creates a new target to better judge trade quality.

Note that every transformation introduces some bias and distortion. This is intentional. Our goal is practical, not theoretical. We care very little about lemmas and proofs. The measure of our success is in the performance of the resulting trading system.

NORMALIZATION

Normalization and standardization are terms that have several meanings and are commonly used interchangeably. One meaning is re-scaling so that the resulting minimum and maximum values are consistent. Another is adjusting the distribution so that the resulting distribution is closer to some reference distribution—such as Uniform, Binary, or Normal.

Standardizing indicators so that the ranges are consistent is a requirement if the indicators will be used as inputs to some modeling methods, such as neural networks. Standardizing is helpful when a template is being used to evaluate indicators or to combine them, as will be our use.

Mathematically, a transformation is the application of a transfer function—a mapping from one set to another set. Think of the RSI indicator applied to closing prices as the input set; and the percent rank of those RSI values as the output set. To be a function, there must be one and only one output value for a given input value.

Some functions are invertible, meaning that the unique input values can be determined from their associated output values. This is a valuable feature in some applications, but is relatively unimportant to trading systems. Many of the transformations we will be using are not invertible—that is, once the transformation has been applied, the original values cannot be recovered.

Some transformations summarize or categorize the input values into a

Transformations

smaller number of output values. While this can be useful, some relationship information is lost, and the transformation is not invertible.

A generally desirable feature of transformations, and one that is important to us, is that the order of the data be preserved. That is, given any two data points, the relationship of their magnitude is the same both before and after the transformation.

RE-SCALING

The simplest transformation is re-scaling. Temperature conversion is an example of re-scaling. Re-scaling is a linear transformation that keeps the proportional distances between any two data points unchanged.

The input data has a minimum and maximum, and the output data has a minimum and maximum. In some cases, the minimum and maximum values are be defined by the indicator, in others they are computed from the sample of data being transformed.

Figure 5.4 illustrates how a set of data from a hypothetical indicator that has data in the range of 30 to 80, are transformed into an equivalent set of data with values from 0 to 100.

Call the two data sets *input* and *output*. Figure 5.4 shows the ranges of the two sets. Assume the latest value of the input set, call it *InputPt*, is 45. The re-scaling transformation computes the corresponding value in the output set.

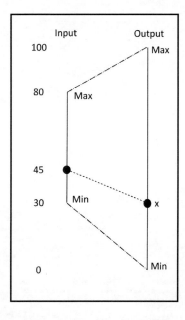

Figure 5.4 - Re-scaling transformation

The arithmetic for the transformation of a value of 45 in the input set to its corresponding value in the output set, 30, is shown by the ratios:

$$\frac{InputPt - InputMin}{InputMax - InputMin} = \frac{OutputPt - OutputMin}{OutputMax - OutputMin}$$

$$\frac{45 - 30}{80 - 30} = \frac{x - 0}{100 - 0}$$

$$x = 30$$

Rearranging terms, the formula for the re-scaling transformation is:

$$OutputPt = (InputPt - InputMin) * \frac{OutputRange}{InputRange}$$

$$OutputPt = (45 - 30) * \frac{100}{50}$$

$$OutputPt = 30$$

Transfer functions are often described in graphical form. Figure 5.5 is the graph of the re-scaling transformation.

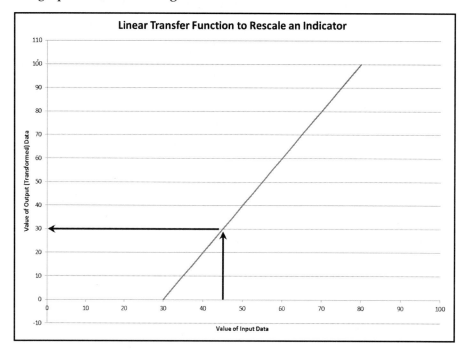

Figure 5.5 - Linear Transfer Function to Re-scale an Indicator

The input is on the horizontal axis. The function is defined for values between 30 and 80, and undefined for all other values. The output is on the vertical axis. The range of output values is 0 to 100. The straight

Transformations

line connecting the points (30, 0) and (80, 100) shows that the function is linear. To interpret a graph and use it to calculate the transformed value of an input value:
1. Locate the input value, 45, on the horizontal axis.
2. Go straight up (or down) to contact the line defining the function.
3. Go straight left (or right) to the vertical axis.
4. Read the output value, 30, from the vertical axis' scale.

To be a true function in the mathematical sense, and to be of value as a transformation for an indicator in a trading system, there must be one and only one output value for every defined input value. That means the vertical line extending from the horizontal axis will intersect the line defining the function in exactly one place.

Since we have decided to have our indicators bounded between 0 and 100, the minimum output value is 0 and the maximum output value is 100. This will be true for all of the transfer functions that fit the standard format we chose. The range of the input is 30 to 80. Values less than 30 and greater than 80 are not defined.

Listing 5.1 shows the AmiBroker code to re-scale a data series, the 14 day CCI, and Figure 5.6 the resulting chart. The CCI function, which is unbounded and can reach values between -300 to 300 and beyond, is rescaled into the range 0 to 100. InputMin and InputMax are calculated to be the maximum CCI values that occurred in the previous 100 days.

```
//    Rescale.afl
//
function rescale( p, inputmin, inputmax, Outputmin, Outputmax )
{
    Outputrange = Outputmax - Outputmin;
    inputrange = inputmax - inputmin;
    x = ( p - inputmin ) * ( Outputrange / inputrange );
    return x;
}

Cc = CCI( 14 );
Plot( Cc, "CCI", colorRed, styleLine | styleDots
        | styleOwnScale, -300, 300 );
CcLookback = 100;
inputmin = LLV( Cc, CcLookback );
inputmax = HHV( Cc, Cclookback );
Ccrescaled = rescale( Cc, inputmin, inputmax, 0, 100 );
Plot( Ccrescaled, "CCI Rescaled", colorBlue, styleLine | styleThick
        | styleOwnScale, -300, 300 );
Plot( 0, "", colorBlue, styleLine | styleOwnScale, -300, 300 );
Plot( 100, "", colorBlue, styleLine | styleOwnScale, -300, 300 );

////////////////    end    ////////////////
```

Listing 5.1 - Re-scale a Series

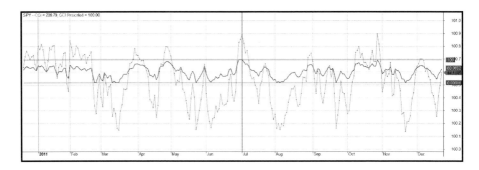

Figure 5.6 - CCI and Re-scaled CCI

Z Transformation

Data are re-scaled to a new mean and standard deviation.

μ_y = The new desired mean of the output.

σ_y = The new desired standard deviation of the output.

μ_x = The computed mean of the input (for some length, l).

σ_x = The computed standard deviation of the input (for length l).

$$y_i = \frac{x_i - \mu_x}{\sigma_x} * \sigma_y + \mu_y$$

Often, $\mu_y = 0$ and $\sigma_y = 1$

So $y_i = \dfrac{x - \mu_x}{\sigma_x}$, z – score.

The z score transformation normalizes the indicator to mean of 0.0 and standard deviation of 1.0, creating a Normal-like distribution almost without regard for the original distribution of the indicator. It is a good transformation to use to emphasize the extreme data points. (This method of normalization is a necessary preprocessing step for data that will be analyzed using principle component analysis.)

```
//   AmiBroker code for z transformation

function zscore( p, Length )
{
//   Transforms input data p
//   to z-score with mean 0 and
//   standard deviation 1.
    zs = ( p - MA( p, Length ) ) / StDev( p, Length );
    return zs;
}
```

```
//   Use of zscore function
z = zscore( C, 9 );
```

Listing 5.2 -- Z Transformation

Out-of-range Values

The input set is a sample—a subset of a population or universe. In some cases, we are certain what the minimum and maximum values will be. RSI, for example, is always between 0 and 100. In other cases, we cannot be certain. Values of %B, for example, are concentrated between 0 and 100, but can be less than 0 or greater than 100. The transformation of range must be able to assign a consistent, accurate, deterministic output value corresponding to any input value, even those outside the expected range. Figure 5.7 shows where this is important.

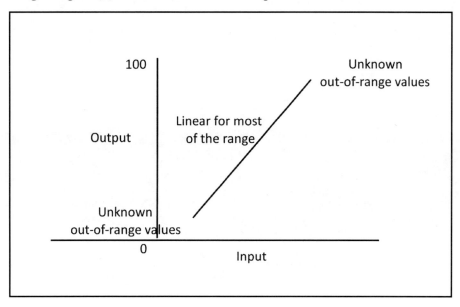

Figure 5.7 -- Anticipating Out-of-Range Values

There are three alternative methods to do this.

Omit or Ignore

All of the price data used in a trading system has already been aligned to a common series, such as SPY, to insure there is a data bar and a price for every period of trading. As indicators are being computed, every bar must have an indicator value that can be interpreted reasonably, even if it is a default value or a copy of the value from the previous bar. Ignoring or omitting is not a reasonable option.

CLIP OR WINSORIZE

Winsorizing is the process of replacing every value that is too low and out of range by the lower limit defined for the output range, and replacing every value that is too high and out of range by the upper limit defined by the output range. All of the left tail data points are given the same low value; and all of the right tail data points are given the same high value. There are drawbacks to Winsorizing that make it a poor choice for our application:
- It fails to preserve the order of the data.
- It fails to distinguish between data points at the extremes—precisely where overbought and oversold are most likely to be identified.

SQUASH—THE LOGISTIC FUNCTION

This transformation maintains as much of the linear relationship (the long straight portion of the transfer function before it begins curving) as possible, while accommodating unknown out-of-range values, however extreme they may be.

While it would be possible to create a function that mated long, tapered sections on the right and left with the linear section in the center, a better solution is to use the logistic function.

The logistic function is one of a class of functions called *sigmoid*. Sigmoid functions vary in the output range (some are -1 to +1, others 0 to 1) and in the steepness of the transition from lower to upper extreme.

The formula for the version of the logistic function we will use is:

$$y(t) = \frac{1}{1+e^{-t}}$$

The value of the logistic function for input values ranging from -10 to +10 was computed in Excel. As defined, the output values of y range from 0 to 1. Multiplying by 100 expands that to our standard range of 0 to 100. Note how the values near the center of the input range, which is near zero, are given a linear transformation, while extreme input values are squashed into the output range. The length of the section that is linear is about six standard deviations. Figure 5.8 shows the plot of the logistic function.

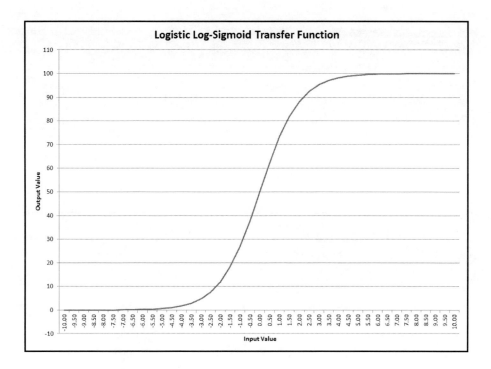

Figure 5.8 -- Logistic function

SOFTMAX FUNCTION

There is another sigmoid function that allows control over how much of the transformation is linear. It is called SoftMax, and it is closely related to the logistic function shown in Figure 5.8. In the logistic function, the section that is linear is six standard deviations—three on either side of the mean. In softmax, the parameter lambda specifies how many standard deviations are linear. The equation is similar to the logistic function, but the mean and standard deviation of the input data sets are used to control what portion of the transformation is linear. The formula, in two pieces for ease of understanding, is:

$$a = \frac{x - \mu}{\frac{\lambda \sigma}{2 \pi}}$$

$$y(t) = \frac{1}{1 + e^{-a}}$$

where:

$e = 2.71828$

π = 3.14159

µ = mean of input data, for a given length *l*

σ = standard deviation of input data for a given length *l*

λ = the number of standard deviations to be kept linear

Figure 5.9 shows the plot of the softmax function for values of lambda of 12 (the steepest), 6, and 3. For each, mean is set to 0 and standard deviation to 1. Note that the logistic function shown in Figure 5.9 and the softmax function when lambda is about 6 (6.283, or 2 times pi, to be more precise) are identical.

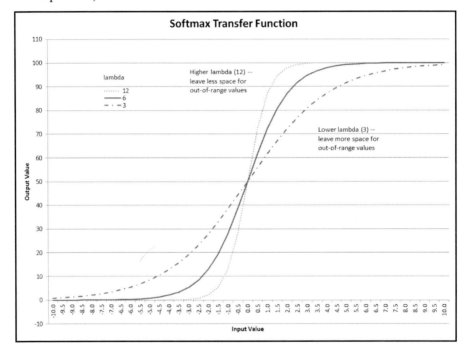

Figure 5.9 -- Softmax function

The SoftMax function accepts data of any range of values as input, normalizes it so that it is centered around zero, and performs the transformation.

The higher the confidence that all the input data will fall into the range observed in the sample, the higher the portion of the output range can be covered by the linear portion of the function, and the higher the value of lambda can be.

The lower the confidence that all the input data will fall into the range observed in the sample, the more space should be allowed for out-of-range data, and the lower the value of lambda should be.

Transformations

This transformation handles out-of-range data, and has limited capability to change the distribution of the in-range portion. Changing from linear to some other distribution can be handled through an additional transformation applied to the output of this one.

Listing 5.3 gives the AmiBroker code to define and apply the SoftMax function.

```
//   SoftmaxFunction.afl
//

//   User defined functions

function Softmax( p, Lambda, Length )
{
//   p -- price or data series
//   lambda -- linearity parameter
//   length -- lookback length

    pi = 3.14159;
    e = 2.71828;

    m = MA( p, Length );
    s = StDev( p, Length );
    a = ( p - m ) / ( ( Lambda * s ) / ( 2 * pi ) );
    y = 1 / ( 1 + e ^ ( -a ) );
    return 100 * y;
}

//   Parameters

RSILength = 14;
Lambda = 6;
SMLookback = 121;

//   Indicators

p = RSI( RSILength );
sm = softmax( p, Lambda, SMLookback );

//   Plots

//Plot( C, "C", colorBlack, styleCandle );
Plot( p, "RSI(14)", colorBlue, styleLine | styleThick
    | styleLeftAxisScale );
Plot( sm, "SoftMax", colorRed, styleDots | styleLeftAxisScale );

////////////////// end //////////////////
```

Figure 5.3 -- SoftMax

Figure 5.10 shows the application of SoftMax to the indicator RSI(14) applied to daily closing prices. Using a 14 period RSI produces values that are near the center of RSI's 0 to 100 range—the heavy solid line in the figure. Transforming the RSI values using SoftMax with lambda of

6 and a 121 day lookback period spreads the output values toward the extremes—the thinner line with dots.

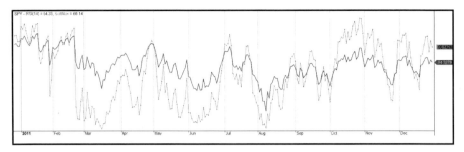

Figure 5.10 -- SoftMax applied to RSI(14)

OTHER SIGMOID FUNCTIONS

HYPERBOLIC TANGENT

The hyperbolic tangent is defined as:

$$\tanh(x) = \frac{e^x - e^{-x}}{e^x + e^{-x}}$$
$$= \frac{e^{2x} - 1}{e^{2x} + 1}$$

It is an excellent, and parameter-free, transformation that accepts inputs of any value, positive or negative, and transforms to the range -1 to +1. An input of 0.0 is transformed to 0.0. Tanh is often used as the second transformation after the raw data is normalized with a z transformation to center it on zero with a standard deviation of 1.0. If the desired output range is 0 to 100, use linear re-scaling—add 1, divide by 2, then multiply by 100.

FISHER TRANSFORMATION

The Fisher transformation is defined as:

$$z = \frac{1}{2} * \ln\left(\frac{1+r}{1-r}\right)$$
$$= \operatorname{arctanh}(r)$$

The Fisher transformation is defined for input values in the range of -1 to +1.

The inverse Fisher transform is defined as:

$$r = \frac{e^{2x} - 1}{e^{2x} + 1}$$
$$= \tanh(z)$$

It is a sigmoid transfer function exactly the same as the hyperbolic tangent. Both tanh and inverse Fisher can be achieved using the SoftMax function with lambda equal to 3.14159.

STATIONARITY

For a time series to be stationary means that it is stable and consistent over time. When any metric, such as the mean or the standard deviation, changes from one period to another, the data is non-stationary with respect to that metric. Whether stationarity is important depends on the analysis being applied. But, in general, removing non-stationarities eases analysis and improves usefulness as an indicator. Two transformations that improve stationarity are differencing and detrending.

Differencing. Differencing is a transformation whose output is the difference between data points. If the magnitude of the data changes significantly through its range, dividing the difference by the price to produce a percent change is often helpful.

Detrending. All time series can be decomposed into two or more related series. One easy and useful method to separate the series is detrending. Compute a moving average of the raw data, then subtract that moving average from the raw data itself. The moving average represents the lower frequency component, while the residuals of the subtraction contain the higher frequency components. Engineers describe this process as *filtering*, the moving average as *a filter*, and the portion of the data remaining after the filter has been applied the *band pass of the filter*. Extensive discussion of filters, their features, and their flaws is beyond the scope of this book.

INDICATORS DEMYSTIFIED

Technical analysis indicators are transformations. They are commonly applied to a portion of a series of prices within a sliding window having some lookback length, resulting in a single numeric value that is associated with the final price in that window. The length of the window is usually fixed at some number of data points or price bars. Think of a window sliding along the data series. The indicator analyzes the data visible through the window, assigning a value to the final position visible.

To be useful in live trading, only data known at the time of the calculation can be used. But peeking into the future is valuable as a research tool in analyzing the potential of an indicator. (Use of data more recent than the final value of the sliding window is known as a *future leak*.)

Application of indicators is not restricted to prices. They can be applied to almost any data series—although some calculations are defined only for certain conditions, such as requiring that all data points be positive. In particular, indicators can be applied to data series resulting from other indicators. This is valuable to change the distribution to emphasize or de-emphasize particular ranges of the indicator.

There is nothing sacred about any transformation or any indicator. The only reason to use an indicator is that it works. Do not assume that an indicator has special properties just because it is a classic in some sense. The method of calculation, length of lookback window, transformation technique, and level of action are all open to modification. By adjusting each of those, most indicators can be configured to give equivalent results. Avoid indicator envy. Pick the one or two indicators you like and have found accurate for your trading systems, and use them. The proof of effectiveness, or lack thereof, comes during the validation phase of system development. If the walk forward out-of-sample results are good, then the transformation, indicator, and rules are good—that is all that matters.

Descriptions, formulas, and plots of a variety of indicators follow. Several analysis techniques are described. Application of a specific technique to a specific indicator is meant to be an educational illustration, not necessarily a recommendation.

Moving Averages

Mean reversion systems begin with identification of the mean—the moving average.

A moving average represents the mean of the data in a sliding window. Although the value of that moving average is associated with the final data point in the window, it measures the mean at an earlier point in the window. The time between the final data point and the point the mean represents is the lag. For a simple moving average, the lag is half the window length. For an exponential moving average, the lag is about half the length parameter.

A moving average cannot detect a change in trend sooner than its lag.

Transformations

When a 200 day moving average changes from sloping upward to sloping downward, it is responding to a change that occurred 100 days ago. In order to be valuable as a predictor, the new trend must last a minimum of 200 days—100 days to recognize the new trend, however many days taking advantage of the new trend, and 100 days while the trend has changed but before the moving average confirms that. To be profitable, by the definition of the moving average, the holding period must be longer than twice the lag.

The moving average is valuable to indicate the mean to which prices will revert. And perhaps, as a filter to identify periods when either long or short positions should not be taken. But it is terrible as an indicator giving buy and sell signals.

We can test the effectiveness of using a 200 day moving average—or any other length—to signal buy and sell. Details of that test are shown in Chapter 9, Systems.

Moving Average Crossover

Traditional technical analysis wisdom suggests that the crossing of two moving averages is important. Commonly used moving averages include simple, exponential, adaptive, and weighted. These differ in the number of data points used and the relative weight each point receives. The results differ in how closely the average tracks the raw data, how much lag there is, and how smooth they are. For some applications, reducing the amount of lag is very important, while others make use of the lag.

The definitions are well known and the functions are built into all trading system development platforms.

We begin by testing the predictive value of moving averages over a short time horizon. Is a moving average crossover system profitable? The program shown in Listing 5.4 varies the type, length, and crossover direction of two moving averages, both based on closing price. Enter a long trade on the cross, exit on the close 1, 3, or 5 days later.

```
//    MovingAverageCrossover.afl
//
//    Test the crossing of two moving averages.
//

SetOption( "InitialEquity", 100000 );
MaxPos = 1;
SetOption( "MaxOpenPositions", MaxPos );
SetPositionSize( 10000, spsValue );
SetOption( "ExtraColumnsLocation", 1 );
```

```
MA1Type = 1; //Optimize("MA1Type",4,1,4,1);
MA2Type = 1; //Optimize("MA2Type",1,1,4,1);
MA1Length = Optimize( "MA1Length", 15, 1, 50, 1 );
MA2Length = Optimize( "MA2Length", 9, 1, 50, 1 );
HoldDays = 1; //Optimize("HoldDays",1,1,5,2);

switch ( MA1Type )
{
case 1:
//   Simple moving average
    MA1 = MA( C, MA1Length );
    break;
case 2:
//   Exponential moving average
    MA1 = EMA( C, MA1Length );
    break;
case 3:
//   Adaptive moving average
    MA1 = DEMA( C, MA1Length );
    break;
case 4:
//   Weighted moving average
    MA1 = WMA( C, MA1Length );
    break;
}

switch ( MA2Type )
{
case 1:
//   Simple moving average
    MA2 = MA( C, MA2Length );
    break;
case 2:
//   Exponential moving average
    MA2 = EMA( C, MA2Length );
    break;
case 3:
//   Adaptive moving average
    MA2 = DEMA( C, MA2Length );
    break;
case 4:
//   Weighted moving average
    MA2 = WMA( C, MA2Length );
    break;
}

Buy = Cross( MA1, MA2 );
Sell = BarsSince( Buy ) >= HoldDays;

///////////////   end   ///////////////
```

Listing 5.4 -- Moving Average Crossover

The variables named MA1Type and MA2Type allow the program to choose the type of moving average to use. The variables MA1Length and MA2Length allow the program to choose the number of bars to use for the moving average. The program issues a buy signal when MA1 crosses up through MA2. Since there is no forced relationship between the two lengths, the crossing could be either of a fast moving average up through a slow moving average (the traditional trend following entry signal) or of a slow moving average crossing up through a fast moving average (a mean reversion entry signal). By setting both types to 1, the system is the crossing of two simple moving averages. The results that are best will depend on each trader's subjective judgement. But none of the most profitable or least risky combinations are trend following.

I found none of the alternatives—any type of moving average, any lengths, trend following or mean reversion—to be worth further evaluation. Try this program yourself on tradable issues of your choice, evaluated by the objective function of your choice.

DETRENDED PRICE OSCILLATOR, INCLUDING POSITION IN RANGE

Subtracting the moving average from the raw price leaves the high frequency residual—a detrended series called the detrended price oscillator.

Listing 5.5 shows the AmiBroker program to compute and plot a detrended price oscillator.

```
//   DetrendedPriceOscillator.afl
//
//   Compute and plot
//   the high frequency residual after detrending.
//

MA1Length = Param( "MA1Length", 20, 1, 50, 1 );

MA1 = MA( C, MA1Length );

DPO = ( C - MA1 ) / MA1;

Plot( C, "C", colorBlack, styleCandle );
Plot( MA1, "MA1", colorBlue, styleLine | styleThick );
Plot ( DPO, "DPO", colorGreen, styleLine | styleOwnScale, -0.05, 0.05 );
Plot ( 0, "", colorGreen, styleDots | styleOwnScale, -0.05, 0.05 );

///////////////   end   ///////////////////
```

Listing 5.5 -- Detrended Price Oscillator

Figure 5.11 shows the plot. The heavy line following the prices is a 20 day simple moving average. The lighter line that is oscillating above and below the horizontal dotted "zero" line is the residual from subtracting the moving average from the closing price. Note that the residual line is above the zero line when the price is above the moving average, crossing the dotted zero line when the moving average is crossing the price, and below when the price is below the moving average. The crossing of the residual with the zero would be the signal for a traditional trend following entry to a long position—the crossing of the closing price above its simple moving average. However, using the program in the previous section, we found that was not a high profit, low risk system.

Figure 5.11 - Plot of Detrended Price Oscillator

The mean reversion alternative is to enter long positions when the residual is below the zero line—oversold—anticipating that prices will revert to the mean. The length of the moving average has been made a Param, so you can adjust its length in the plot pane and see the change immediately. The residual has been divided by the moving average to standardize the magnitude of the deviation as a percentage.

One way to design a trading system based on this is to manually adjust the length, searching for a some consistent degree of oversold. But the deviation at the valleys is great. They are in the range of one or two percent for long periods of time, punctuated by occasional deviations of five percent or more. Some transformation is needed to help identify valleys.

Position-in-range is a transformation that might help. Listing 5.6 shows the AmiBroker code to implement the PIR function. Use the Param feature to investigate parameter settings that might identify profitable trades.

```
//    DetrendedPriceOscillator_WithPIR.afl
//
//    Compute and plot
// the high frequency residual after detrending.
//

function PIR( p, Length )
{
//    compute the position in range of series p,
//    relative to its position over
//    a lookback window of Length.
```

Transformations

```
    ll = LLV( p, Length );
    hh = HHV( p, Length );
    return( 100*( p - ll ) / ( hh - ll ) );
}

MA1Length = Param( "MA1Length", 3, 1, 50, 1 );
PIRLength = Param( "PIRLength", 12, 2, 30, 1 );

MA1 = MA( C, MA1Length );

DPO = ( C - MA1 ) / MA1;
PIRDPO = PIR( DPO, PIRLength );

Plot( C, "C", colorBlack, styleCandle );
Plot( MA1, "MA1", colorBlue, styleLine | styleThick );
Plot ( PIRDPO, "PIRDPO", colorGreen, styleLine
          | styleOwnScale, 0, 100 );

////////////////   end   ////////////////
```

Listing 5.6 - Position-in-Range of Detrended Oscillator

Figure 5.12 shows the plot of the detrended price oscillator, transformed by the position-in-range function. Call it PIRDPO.

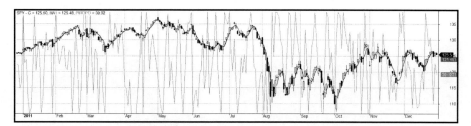

Figure 5.12 - Plot of DPO transformed by PIR

Convert this to a trading system by adding a signal to buy when PIRDPO is at a low level. Listing 5.7 shows the AmiBroker code.

```
//   DetrendedPriceOscillator_System.afl
//
//   Compute and plot
// the high frequency residual after detrending.
//

function PIR( p, Length )
{
// compute the position in range of series p,
// relative to its position over
// a lookback window of Length.
    ll = LLV( p, Length );
    hh = HHV( p, Length );
    return( 100*( p - ll ) / ( hh - ll ) );
}

MA1Length = Optimize( "MA1Length", 3, 1, 50, 1 );
PIRLength = Optimize( "PIRLength", 12, 2, 30, 1 );
```

```
        BuyLevel = Optimize( "BuyLevel", 10, 1, 30, 1 );

        MA1 = MA( C, MA1Length );

        DPO = ( C - MA1 ) / MA1;
        PIRDPO = PIR( DPO, PIRLength );

        Buy = PIRDPO < BuyLevel;
        Sell = BarsSince( Buy ) >= 3;

        Plot( C, "C", colorBlack, styleCandle );
        Plot( MA1, "MA1", colorBlue, styleLine | styleThick );
        Plot ( PIRDPO, "PIRDPO", colorGreen, styleLine
                | styleOwnScale, 0, 100 );

//////////////// end ////////////////
```

Listing 5.7 -- Trading System Based on PIRDPO

DV2, INCLUDING RANK TRANSFORMATION

Credit for describing DV2 goes to David Varadi of CSS Analytics. The indicator compares the close with the average of the high and low.

$$y = \left(\frac{C}{0.5 * (H+L)}\right) - 1$$
$$DV(2) = MA(y, 2)$$

where

MA is a simple moving average

2 is the length of the lookback window

Since the closing price is always between the high and the low, the maximum and minimum values of y are usually between -1 and +1, but can extend beyond either. Figure 5.13 shows the plot of DV(2) before any transformations have been applied to it.

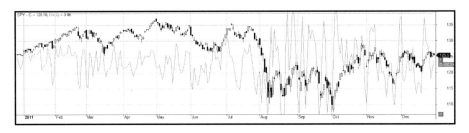

Figure 5.13 - DV(2)

Figure 5.14 shows the distribution of DV(2) for SPY for the period 1/1/1999 to 1/1/2012.

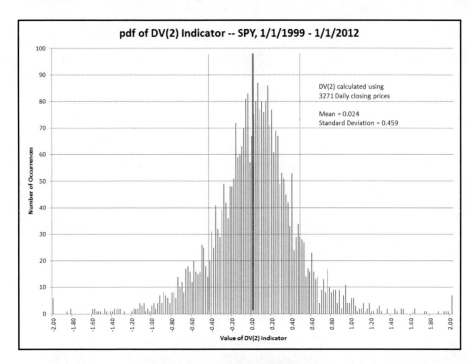

Figure 5.14 - pdf of DV(2) Indicator

For traders who will be entering at the close on the day the buy signal is generated, and exiting on the close the day the sell signal is generated, the quality of the trade is the percentage change from close to close for the period ahead. In order to examine performance of an indicator, we compute the value of the indicator as of the close for every daily bar, and also the change in price ahead 1, 2, and 3 days. Using the PercentRank transformation function to linearize the distribution of the DV indicator, naming the new indicator PRDV, we create a table using AmiBroker's Explore functionality, then export the table in csv format.

Listing 5.8 shows the AmiBroker code that Explore uses to generate the table of results.

```
//   DVGainAhead.afl
//
//   DV indicator
//   Per David Varadi
//
```

```
function DV( MALength )
{
    y = ( C / ( 0.5 * ( H + L ) ) ) - 1;
    return ( 100 * MA( y, MALength ) );
}

DVIndic = DV( 2 );
PRDV = PercentRank( DVIndic, 100 );
GainAhead1 = ( Ref( C, 1 ) - C ) / C;
GainAhead2 = ( Ref( C, 2 ) - C ) / C;
GainAhead3 = ( Ref( C, 3 ) - C ) / C;

Plot( C, "C", colorBlack, styleCandle );
Plot( DVIndic, "DV(2)", colorGreen, styleLine | styleOwnScale, -1, 1 );
Plot( -1, "", colorGreen, styleLine | styleOwnScale, -1, 1 );
Plot( 1, "", colorGreen, styleLine | styleOwnScale, -1, 1 );

// Use Explore to create a table

Filter = 1;
AddColumn( DVIndic, "DV", 10.6 );
AddColumn( PRDV, "PRDV", 10.2 );
AddColumn( GainAhead1, "GainAhead1", 10.6 );
AddColumn( GainAhead2, "GainAhead2", 10.6 );
AddColumn( GainAhead3, "GainAhead3", 10.6 );

///////////// end /////////////////
```

Listing 5.8 -- Generate DV2 Table

Figure 5.15 shows the distribution of the new indicator, PRDV. PRDV is DV after the PercentRank transformation has been applied. It should be no surprise that it is approximately a uniform distribution. The fact that it is not perfectly uniform is due to the selection of the parameter used by PercentRank. The value of 100 meant that each bar was ranked within a sliding window of 100 days. Using a different window length will result in a different distribution, longer lookback lengths producing more nearly uniform results. In order to be perfectly uniform, knowledge of future data would be required.

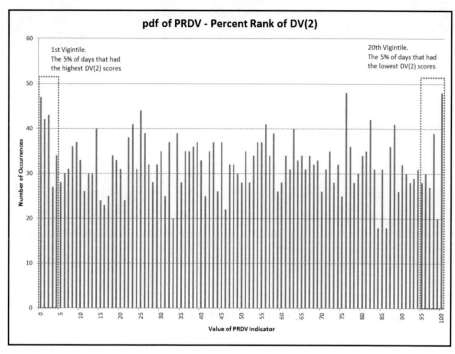

Figure 5.15 - pdf of Percent Rank of DV(2)

PRDV will be useful if there is a relationship between the value of PRDV and the future price change.

Using Excel, open the csv file. Sort the 3271 data elements by PRDV value. Create 20 equal-sized groups. (Ten equal-sized groups are deciles, twenty are vigintiles.) The first vigintile contains all the data points with PRDV values 0 through 4. For each vigintile, compute the average change in closing price 1, 2, and 3 days ahead.

Figure 5.16 shows the relationship. There are three lines—1 day ahead (dotted line), 2 days (small dashes), and 3 days (long dashes). The y axis is percentage change for the entire period (not per day). The x axis is vigintile number with lowest scoring 5% on the left, highest 5% on the right. There is a linear regression for each of the 20 point data series.

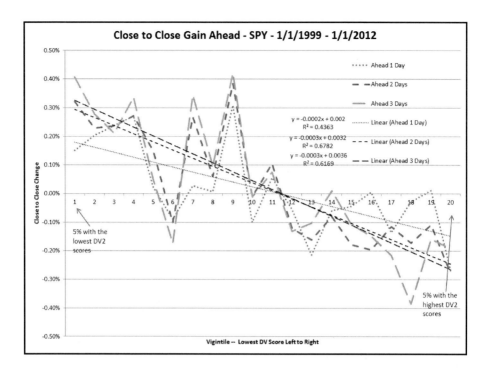

Figure 5.16 - PRDV in Vigintiles

All three of the lines have the characteristics we are looking for. Low indicator scores correspond to positive future price changes, high scores correspond to negative future price changes, the relationship between score and future price change is reasonably linear, and the results for all three days ahead are consistent. Based only on this analysis, two trading systems are suggested:

1. Buy at the close of every day when the score of DV2 is in the bottom 20% (four vigintiles).
   Sell by the end of the third day.
2. Short at the close of every day when the score of DV2 is in the top 30% (six vigintiles).
   Cover by the end of the third day.

All of the analysis so far has been exploratory, and these results are entirely in-sample. Additional system development work must be done to determine the stability of the relationship over time. But DV2 appears to be a promising indicator.

RANK TRANSFORMATION

As demonstrated in the previous section, rank transformation sorts the data in the lookback window into ascending order and assigns each data point the value of its position in the sorted order. Rank transformation is a linearizing transformation. The output of rank transformation is approximately a uniform distribution.

Depending on implementation, ties among input data will either:
- All be given the average of the ranks of the tied points.
- Each be given a unique rank, with all the tied points adjacent to each other in the output.

PercentRank is a rank transformation function available as a built-in function for both Excel and AmiBroker.

Diffusion Index

Diffusion indexes are composites formed from many individual results. An indicator is formed from combining the contributing values in some way.

A common diffusion index is the advance-decline index. The value contributed by each individual is whether the most recent closing price is higher or lower than the previous close.

Interpretation and effectiveness depends on:
- Membership.
- Time horizon being forecast.
- What is being forecast.

Membership is extremely sensitive to selection bias and survivorship bias. Take the advance-decline index as an example.

Selection relates to which categories are included—choices include: common stocks, preferred stocks, foreign stocks, warrants, exchange traded funds, constituents of an index.

Survivorship relates to the list of issues being used and the time period used. If the membership is constituents of the S&P 500 index, then a new membership list is required whenever an issue leaves or enters the index. The database used must include data for delisted issues.

Rough diffusion indexes can be created from reported broad market indicators. The number of advancing issues, declining issues, and unchanged issues is reported for the NYSE, NASDAQ, and AMEX. A diffusion index for the NYSE is computed as:

$$DiffIndex = \frac{AdvIss}{AdvIss + DecIss}$$

Historical daily data for NYSE, NASDAQ, and other exchanges, for advancing issues, declining issues, new highs, and other statistics, is widely available. Provided that we are willing to accept whatever membership definition the reporting agency used, we can compute diffusion indexes based on those values..

The AmiBroker program to compute an advance-decline diffusion index based on NYSE and compute future performance is shown in Listing 5.9. The forecast horizon has been set to one day.

This program is more useful than it might appear. Change the indicator and / or the forecast horizon to create a table of future performance.

```
//   IndicatorAnalysisTemplate.afl
//
//   Template to create a set of columnar data
//   giving indicator value and performance
//   over the forecast horizon.
//
//   Use AmiBroker's Analysis > Explore.
//   Export the columns of data to Excel for analysis
//

SetForeign( "$NYA" );
Plot( C, "C", colorBlack, styleCandle );

ADV = Foreign( "#NYSEADV", "C" );
DEC = Foreign( "#NYSEDEC", "C" );
AdvDecLine = 100 * ADV / ( ADV + DEC );

Plot( AdvDecLine, "ADLine", colorRed, styleLine | styleOwnScale );

//   The indicator
Indic = AdvDecLine;

//   The forecast horizon
Horizon = 1;

//   Examine all data

for ( i = 1;i < BarCount;i++ )
{
//   Entry is C[i]
//   Exit is C[i+Horizon]

    AE[i]  = 0;   //  Adverse Excursion
    MAE[i] = 0;   //  Maximum Adverse Excursion
    FE[i]  = 0;   //  Favorable Excursion
    MFE[i] = 0;   //  Maximum Favorable Excursion
    EQ[i]  = 0;   //  Equity, relative to entry
    MEQ[i] = 0;   //  Maximum Equity
    DD[i]  = 0;   //  Drawdown
    MDD[i] = 0;   //  Maximum Drawdown
```

```
        for ( j = 1; j <= Horizon && i + j < BarCount; j++ )
        {
            EQ[i] = ( C[i+j] - C[i] ) / C[i];
            MEQ[i] = Max( EQ[i], MEQ[i] );
            AE[i] = ( C[i] - L[i+j] ) / C[i];
            MAE[i] = Max( AE[i], MAE[i] );
            FE[i] = ( H[i+j] - C[i] ) / C[i];
            MFE[i] = Max( FE[i], MFE[i] );
            DD[i] = MEQ[i] - EQ[i];
            MDD[i] = Max( DD[i], MDD[i] );
        }
}

//  Columns from Explore

Filter = 1;

AddColumn( O, "Open", 10.6 );
AddColumn( H, "High", 10.6 );
AddColumn( L, "Low", 10.6 );
AddColumn( C, "Close", 10.6 );
AddColumn( Indic, "AdvDecLine", 10.6 );
AddColumn( Horizon, "Horizon", 10.0 );
AddColumn( EQ, "Equity", 10.6 );
AddColumn( MEQ, "Max Equity", 10.6 );
AddColumn( DD, "Drawdown", 10.6 );
AddColumn( MDD, "Max DD", 10.6 );
AddColumn( FE, "Fav Excur", 10.6 );
AddColumn( MFE, "Max Fav Excur", 10.6 );
AddColumn( AE, "Adv Excur", 10.6 );
AddColumn( MAE, "Max Adv Excur", 10.6 );

///////////////  end  ///////////////
```

Listing 5.9 -- Future Performance

Figure 5.17 shows a graph of the advance-decline indicator, along with the NYSE Composite Index, NYA.

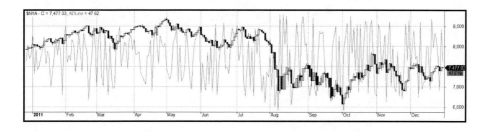

Figure 5.17 - Advance-Decline Diffusion Index

Using AmiBroker's Explore procedure, a set of daily data is produced, exported in csv format, and processed in Excel. Figure 5.18 shows the close to close gain one day ahead for each of the twenty vigintiles of index score. The lowest 15% of index scores appear to have the largest one day gain.

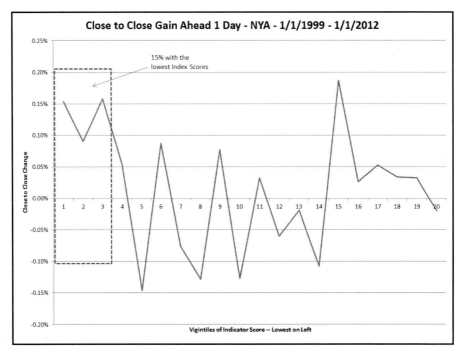

Figure 5.18 - Advance Decline Diffusion Index in Vigintiles

Volatility is high for the vigintile with the lowest index score. Figure 5.19 shows that both maximum favorable excursion and maximum adverse excursion are the highest for the first vigintile.

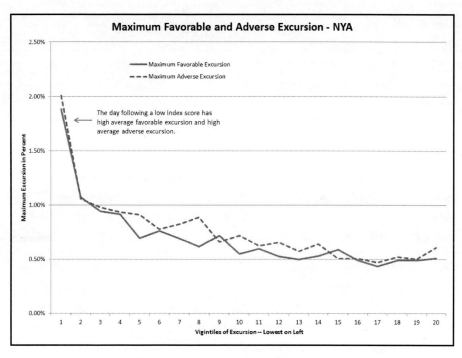

Figure 5.19 - MFE and MAE in vigintiles

This might be an indication that a low percentage of new highs is, on average, a signal to take a long position, but with volatility so high for the very lowest vigintile, it might be prudent to remain flat when the reading is in the lowest 5%. You can implement this by first applying the percentrank transformation, resulting in the new indicator named IndicPR. Then set a minimum limit as well as a maximum limit for taking a long position. The statement will be:

```
Buy = IndicPR < 15 and IndicPR > 5;
```

Note that the distribution is not symmetrical. It may not be equally accurate in predicting long entries and short entries.

We are creating a system that is long and flat. Only the left end of the distribution appears to have potential to forecast long positions. A separate analysis would be carried out for a companion system that trades short and flat; and this indicator does not seem to be a good candidate for inclusion in it.

Be aware that it is easy to create a future leak using this program. As the program is listed, it uses NYSE advancing issues and NYSE declining issues to create the diffusion indicator. When the columnar data is created using AmiBroker's Explore procedure, data is reported for all 3271 days. This data is exported to Excel, where it is analyzed. In Excel, all 3271 data elements are sorted into vigintiles without regard for their time sequence. The data grouping and charts that result represent all 3271 data points. Judging by a visual observation of Figure 5.18, the distribution of the diffusion index appears to be consistent over the entire 13 year period. But it is easy to create indicators that have a concentration of, say, high values in one period of time and low values in another period. Including them all in a single analysis might be misleading.

Vigintile analysis is just the beginning. The proof of whether low values precede profitable long positions will be determined during the validation phase of system development by the out-of-sample results of the walk forward runs.

STOCHASTIC

Stochastic is an indicator of the position of a single data point in relation to the range of values of that data. In its traditional use, the stochastic relates the final price for a period of time—a sliding window with a finite lookback length—to the highest and lowest prices observed in that period.

The calculation is:

$$s_r = y(p,l) = 100 * \frac{p_i - Min(p)}{Max(p) - Min(p)}$$

where:

s_r is the raw stochastic

p is the series of data points being transformed

p_i is the final value

l is the length of the lookback period

$Min(p)$ is the lowest value of p in the lookback period

$Max(p)$ is the highest value of p in the lookback window

When the formula is changed slightly—computing the position relative to the highest value rather than the lowest—the indicator is called Williams %R.

RSI

The misnamed Relative Strength Index, RSI, is credited to J. Welles

Wilder and dates from 1978. Calculation has nothing to do with relative strength. It is based on only a single price from each bar—usually the closing price. The formula was originally and inelegantly expressed as:

$$RSI = 100 - \frac{100}{1+RS}$$

where RS is the ratio of gains to losses. Bar-by-bar, changes in the series are sorted into two series—gains and losses—each of which is smoothed using Wilder's version of an exponential moving average. With a little algebra, the formula for RSI can be simplified, clarified, and made more generally useful.

Let G represent the series of gains and L represent the series of losses, with

$$RS = \frac{G}{L}$$

$$\begin{aligned}
RSI &= 100 - \frac{100}{1+RS} \\
&= \frac{100(1+RS) - 100}{1+RS} \\
&= \frac{100\,RS}{1+RS} \\
&= 100 * \frac{RS}{1+RS} \\
&= 100 * \frac{\frac{G}{L}}{1+\frac{G}{L}} \\
&= 100 * \frac{\frac{G}{L}}{\frac{L+G}{L}} \\
&= 100 * \frac{G}{L+G}
\end{aligned}$$

RSI simplifies to the ratio of Gains to Gains plus Losses. Seen in this form, RSI is akin to a diffusion index—an advance-decline index based on the direction of previous closes.

Wilder's moving average is identically the exponential moving average after adjusting the length parameter.

WilderMovingAverage(p,n) is the same as ExponentialMovingAverage(p,(2n-1)).

Several versions of AmiBroker code follow in sequence, beginning with the standard RSI and concluding with a much more useful custom indicator.

The first, Listing 5.10, begins with the standard, built-in RSI, and demonstrates that it can be replicated using the series of gains and losses. Smoothing is done using the Wilder method. Note that the lookback length variables are defined using Param statements. By right-clicking the pane in which these indicators are plotted, you can vary the lengths and compare them.

```
//   RSI_Custom_V1.afl
//
//   A Modified RSI
//   Replicating built-in RSI indicator
//

RSILengthBuiltIn = Param( "RSILengthBuiltIn", 14, 2, 30, 1 );
RSIBuiltIn = RSI( RSILengthBuiltIn );

RSILengthCustom = Param( "RSILengthCustom", 14, 2, 30, 1 );
UpMove = Max( C - Ref( C, -1 ), 0 );
DnMove = Max( Ref( C, -1 ) - C, 0 );
UpMoveSm = Wilders( UpMove, RSILengthCustom );
DnMoveSm = Wilders( DnMove, RSILengthCustom );
Numer = UpMoveSm;
Denom = UpMoveSm + DnMoveSm;

RSICustom = 100 * IIf( Denom <= 0, 0.5, Numer / Denom );

Plot( C, "C", colorBlack, styleCandle );
Plot( RSICustom, "RSICustom", colorGreen, styleLine
      | styleOwnScale, 0, 100 );
Plot( RSIBuiltIn, "RSIBuiltIn", colorBlue, styleLine
      | styleOwnScale, 0, 100 );

///////////////////   end   ///////////////////
```

Listing 5.10 -- Replicate RSI using Two Series

The next, Listing 5.11, replaces the Wilder Moving Average with the Exponential Moving Average. Note the change in the lookback period necessary so that two plotted lines agree.

```
//   RSI_Custom_V2.afl
//
//   A Modified RSI
//   Using Exponential Moving Average
//

RSILengthBuiltIn = Param( "RSILengthBuiltIn", 14, 2, 30, 1 );
RSIBuiltIn = RSI( RSILengthBuiltIn );

RSILengthCustom = Param( "RSILengthCustom", 27, 2, 40, 1 );
UpMove = Max( C - Ref( C, -1 ), 0 );
```

```
DnMove = Max( Ref( C, -1 ) - C, 0 );
UpMoveSm = EMA( UpMove, RSILengthCustom );
DnMoveSm = EMA( DnMove, RSILengthCustom );
Numer = UpMoveSm;
Denom = UpMoveSm + DnMoveSm;

RSICustom = 100 * IIf( Denom <= 0, 0.5, Numer / Denom );

Plot( C, "C", colorBlack, styleCandle );
Plot( RSICustom, "RSICustom", colorGreen, styleLine
      | styleOwnScale, 0, 100 );
Plot( RSIBuiltIn, "RSIBuiltIn", colorBlue, styleLine
      | styleOwnScale, 0, 100 );

////////////////////  end  ////////////////////////
```

Listing 5.11 -- Replicate using EMA

SIDEBAR: EMA AND LAMBDA

To compute an exponential moving average:
1. Select the weight to be given to the new data value to be added to the average.
2. Initialize the average to the first (oldest) value.
3. For all additional values:
   A. Multiply the new data value by lambda.
   B. Multiply the previous average by 1 - lambda.
   C. Add those two together to give the new value of the average.

A key point is that there is no mention of the number of bars. Just as a small percentage of a well established solera Sherry is very old wine, a small percentage of every exponential moving average is very old data. The number of bars that contribute to an exponential moving average is "all of them." The parameter that controls the smoothing of an exponential moving average is lambda, λ, the weight given to the newest data point. For the convenience of computation and comparison to other moving averages, a pseudo-lookback length, n, that gives an approximate equivalent smoothing and lag is defined:

$$\lambda = \frac{2}{n+1}$$
$$n = \frac{2}{\lambda} - 1$$

n is an artifact. There is no reason it must be an integer. It is lambda that controls the EMA. n is useful in estimating the number of bars in the past that the EMA measures—approximately one-half n.

END OF SIDEBAR

The next, Listing 5.12, demonstrates use of the relationship between n, the lookback length, and lambda, the weighting of the most recent data point.

```
//    RSI_Custom_V3.afl
//
//    A Modified RSI
//    Compute Lookback based on lambda.
//

RSIBuiltin = RSI( 14 );

//    The relationship between lambda and the
//    equivalent length of an exponential
//    moving average is:
//    lambda = 2 / (LB+1)

Lambda = Param( "Lambda", 0.07, 0.01, 0.99, 0.01 );

RSILB = ( 2 / Lambda ) - 1;
//    RSILB will be converted to an integer by
//    the EMA function in this version of the program
//    lambda == 0.07 results in RSILB = 27
//    From other versions of this program,
//    we know that RSI(14) using Wilder is
//    the same as RSI_Custom(27) using EMA.

UpMove = Max( C - Ref( C, -1 ), 0 );
DnMove = Max( Ref( C, -1 ) - C, 0 );
UpMoveSm = EMA( UpMove, RSILB );
DnMoveSm = EMA( DnMove, RSILB );
Numer = UpMoveSm;
Denom = UpMoveSm + DnMoveSm;

RSICustom = 100 * IIf( Denom <= 0, 0.5, Numer / Denom );

Plot( C, "C", colorBlack, styleCandle );
Plot( RSICustom, "RSICustom", colorGreen, styleLine
      | styleOwnScale, 0, 100 );
Plot( RSIBuiltIn, "RSIBuiltIn", colorBlue, styleLine
      | styleOwnScale, 0, 100 );

////////////////////    end    ////////////////////
```

Listing 5.12 -- Demonstrate EMA's Lambda

The next, Listing 5.13, uses looping code to explicitly compute the exponential moving average. This allows use of lambda to specify the weighting of the most recent data added to the average, which removes the requirement that the lookback period be an integral number of bars.

```
//    RSI_Custom_V4.afl
//
//    A Modified RSI
//    Allowing fractional lookback lengths
//
```

Transformations

```
//   The relationship between lambda and the
//   equivalent length of an exponential
//   moving average is:
//   lambda = 2 / (LB+1)
//   A traditional RSI(2) is equivalent to
//   RSI_Custom(3) using EMA.
//   To increase resolution, use lambda rather than LB.

Lambda = Param( "Lambda", 0.50, 0.11, 0.99, 0.01 );

UpMove = Max( C - Ref( C, -1 ), 0 );
DnMove = Max( Ref( C, -1 ) - C, 0 );

//   Initialze arrays
UpMoveSm = C;
DnMoveSm = C;

for ( i = 1;i < BarCount;i++ )
{
    UpMoveSm[i] = Lambda * UpMove[i]
            + ( 1.0 - Lambda ) * UpMoveSM[i-1];
    DnMoveSm[i] = Lambda * DnMove[i]
            + ( 1.0 - Lambda ) * DnMoveSM[i-1];
}

Numer = UpMoveSm;

Denom = UpMoveSm + DnMoveSm;

RSICustom = 100 * IIf( Denom <= 0, 0.5, Numer / Denom );

Plot( C, "C", colorBlack, styleCandle );
Plot( RSICustom, "RSICustom", colorGreen, styleLine
        | styleOwnScale, 0, 100 );

///////////////////  end  ///////////////////
```

Listing 5.13 -- Allow Fractional Lookback Periods

The next, Listing 5.14, uses a user-defined function to implement the custom RSI. Any data series can be passed to the function, as well as any meaningful value of lambda.

```
//   RSI_Custom_V5.afl
//
//   A Modified RSI
//   Allowing fractional lookback lengths
//   coded as a function
//

//   The relationship between lambda and the
//   equivalent length of an exponential
//   moving average is:
//   lambda = 2 / (LB+1)
//   A traditional RSI(2) is equivalent to
//   RSI_Custom(3) using EMA.
//   To increase resolution, use lambda rather than LB.
```

```
function RSI_Custom( p, Lambda )
{
    //  p == series having its RSI computed
    //  lambda == weight given to latest value
    UpMove = Max( p - Ref( p, -1 ), 0 );
    DnMove = Max( Ref( p, -1 ) - p, 0 );

//  Initialze arrays
    UpMoveSm = p;
    DnMoveSm = p;

    for ( i = 1;i < BarCount;i++ )
    {
        UpMoveSm[i] = Lambda * UpMove[i]
            + ( 1.0 - Lambda ) * UpMoveSM[i-1];
        DnMoveSm[i] = Lambda * DnMove[i]
            + ( 1.0 - Lambda ) * DnMoveSM[i-1];
    }

    Numer = UpMoveSm;

    Denom = UpMoveSm + DnMoveSm;

    return ( 100 * IIf( Denom <= 0, 0.5, Numer / Denom ) );
}

Lambda = Param( "Lambda", 0.50, 0.11, 1.0, 0.01 );

RSI_C = RSI_Custom( C, Lambda );

Plot( C, "C", colorBlack, styleCandle );
Plot( RSI_C, "RSICustom", colorGreen, styleLine
        | styleOwnScale, 0, 100 );

////////////////////    end    ////////////////////
```

Listing 5.14 -- Custom RSI Function

The final, Listing 5.15, uses the custom RSI in a mean reversion trading system. Note that lambda is non-linear. Valid range is 0.01 to 1.00. Values of lambda corresponding to short-lookback RSI systems, such as RSI(2), are around 0.60 to 0.80. Lambda equal to 1.00 uses no smoothing. Values above 1.00 are not meaningful. To reduce the sensitivity, this version of the program has a switch that allows specification of the moving average length either in terms of lambda directly, or lambda computed from the lookback length. To replicate RSI(2) using Wilder's moving average, use RSI(3) using exponential moving average, which corresponds to lambda of 0.50.

```
//  RSI_Custom_V6.afl
//
//  Trading system based on a modified RSI.
//  Allowing fractional lookback lengths
//  coded as a function.
```

```
//
//  The relationship between lambda and the
//  equivalent length of an exponential
//  moving average is:
//  lambda = 2 / (LB+1)
//  A traditional RSI(2) is equivalent to
//  RSI_Custom(3) using EMA.
//  To increase resolution, use lambda rather than LB.

OptimizerSetEngine( "cmae" );

SetOption( "Initialequity", 100000 );
MaxPos = 10;
SetOption( "MaxOpenPositions", MaxPos );
SetPositionSize( 10000, spsValue );

SetOption( "ExtraColumnsLocation", 1 );

function RSI_Custom( p, Lambda )
{
    //  p == series having its RSI computed
    //  lambda == weight given to latest value
    UpMove = Max( p - Ref( p, -1 ), 0 );
    DnMove = Max( Ref( p, -1 ) - p, 0 );

//  Initialze arrays
    UpMoveSm = p;
    DnMoveSm = p;

    for ( i = 1;i < BarCount;i++ )
    {
        UpMoveSm[i] = Lambda * UpMove[i]
            + ( 1.0 - Lambda ) * UpMoveSM[i-1];
        DnMoveSm[i] = Lambda * DnMove[i]
            + ( 1.0 - Lambda ) * DnMoveSM[i-1];
    }

    Numer = UpMoveSm;

    Denom = UpMoveSm + DnMoveSm;

    return ( 100 * IIf( Denom <= 0, 0.5, Numer / Denom ) );
}

//  Alternative parameter specification
UseLambdaDirectly = 1;

if ( UseLambdaDirectly )
{
//  The upper limit on lambda is 1.0.
    Lambda = Optimize( "Lambda", 0.46, 0.16, 1.0, 0.01 );
//  0.16 == Lookback = 11 (using EMA) = 6 (using Wilder)
//  1.00 == Lookback = 1 (no smoothing)
}
else
{
//  Lambda is non linear.
//  Having too fine a level of control over
```

```
//  lambda may not be the best technique.
//  This method begins with the Lookback period,
//  and computes the corresponding lambda.
    LBPeriod = Optimize( "LBPeriod", 2, 1.5, 6, 0.25 );
    Lambda = 2 / ( LBPeriod + 1 );
}

RSI_C = RSI_Custom( C, Lambda );

BuyLevel = Optimize( "BuyLevel", 28, 5, 30, 1 );

Buy = RSI_C < BuyLevel;
Sell = RSI_C > 50;


Plot( C, "C", colorBlack, styleCandle );
Plot( RSI_C, "RSICustom", colorGreen, styleLine
     | styleOwnScale, 0, 100 );

/////////////////////   end   /////////////////////
```

Listing 5.15 - Trading System using Custom RSI

We have replicated the traditional RSI precisely, and added significant flexibility to it. But even if the function RSI_Custom did not replicate the traditional RSI, we could simply have given it a name, claimed to have invented a new indicator, claimed trademark protection, and used it. OK, I will not claim trademark—I trust you will not either.

Exotic Indicators

There are an immense number of other indicators. If I omitted your favorite, let me know and I will consider including it in the next edition.

Some indicators that are not covered here that show promise are somewhat exotic:
- Dimension reduction methods, such as Principal Component Analysis.
- Classification methods based on clustering, nearest neighbor analysis, and decision trees.
- Neural networks.

These are not yet available in most retail-level trading system development packages.

Remember that the goal is to have confidence in trading signals, and that confidence comes from the validation phase of system development. If you do use analysis tools outside your development platform, decide early on in the development process how you will validate those systems.

BACK TO THE VERTICAL AXIS

Now that we have examined some indicators, and some techniques for transforming them to better serve our purposes we return to the question: "What do we want plotted on the vertical axis?"

Historically, indicators have been somewhat divided into categories of trending or mean reverting according to each indicator's primary use. To my thinking, that division is both artificial and unnecessary. Every profitable trade depends on a price change—a trend—between the entry to the trade and exit from it. The quality of a trade depends on both the profit gained when the trade is closed and the risk experienced while the trade is open.

The output of your trading system is a sequence of buy and sell signals. The set of trades that result from following those signals have some performance characteristics. They can be summarized by an equity curve for a quick visual analysis. Two aspects are central:
- How fast does the account grow?
- How deep a drawdown will I experience?

Modeling Trading System Performance goes into the details of analyzing the set of trade results to give quantified answers to both those questions. This book focuses on the rules that result in trades.

QUALITY MATTERS

Your system is looking for patterns that precede profitable trades. As with most things we experience, the clarity and certainty of the pattern, and the profitability and risk of the trade, have considerable variability.

For ease of discussion, assume every system can be expressed as testing the level of a numerical indicator. When the level is above some limit, be long; below that level, be flat. The ideal indicator is shown in Figure 5.1 where the highest indicator values precede the highest quality trades— high profit and low risk. The trades, or potential trades, that follow lower readings are lower quality—they have lower profit and / or higher risk.

Where should you set your buy level? As always, there is some subjectivity that you must resolve according to your personal risk preference. But the procedure is the same for everyone. One way is to let your trading system development platform interpret the levels as buy or sell signals, then evaluate the resulting trades.

Alternatively, you can evaluate the indicator independently. For every bar, compute the value of the indicator. Examine the prices over the bars that follow, assigning a quality metric to each bar that scores the trade potential. Sort the results by indicator value, perhaps into decile or vigintile bins. For each bin, evaluate trade potential. An advantage to using this method is that it reveals bad behavior that exists in some indicators. Usually higher indicator values are better, but in some cases performance following the most extreme readings is poor.

Since the systems we are developing hold only a few days, the forecast horizon can be short; 1, 2, or 3 days is probably adequate. If the price does not change in the direction that will give our trades a profit in the first day or two, the signal is not very good for our purposes.

If you will always enter every trade at a given time, such as the close of trading or the open of trading, and always exit every trade at a given time, never taking a profit target, using a trailing exit, or using a maximum loss exit, then a single metric is all that is needed—price change between entry and exit—as used in Figure 5.1.

If you will, or could, enter intra-bar at a limit or at a stop, or exit intra-bar at a limit or at a stop, then the intra-bar high and low prices are important. If you plan to exit using a profit target, the highest price over the next few bars is more important than the closing price some number of bars in the future. Similarly, if you will exit a trade because it has an intra-trade drawdown, the low price becomes very important. Designing a metric that combines risk with profit potential is complex and will depend on your subjective choices. But, in general, high quality trades have high profit potential and low risk, while low quality trades have low profit potential and/or high risk. As Figure 5.20 shows graphically, you want to avoid the shaded areas.

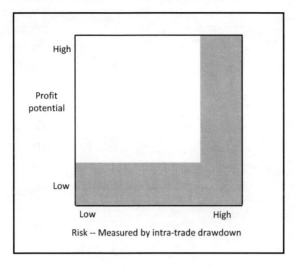

Figure 5.20 -- Trade Quality

It seems reasonable to measure profit potential as: the percentage gain from entry to optimal exit, divided by the number of days exposed; and to measure risk by the percentage loss to extreme drawdown over the entire period exposed. The trade quality metric might be a ratio of the two, checking for a zero divisor and assigning a value of, say, 10 when there is no intra-trade drawdown. The program in Listing 5.8 computes gain, profit potential, and risk, allowing you to analyze them as you prefer.

6

Exits

We are discussing exits early, not because they are necessarily more important than the other components of systems, but because we will use some of the results of the studies of exits to simplify the system development process.

The studies that follow are based on daily closing prices of SPY for the period 1/1/1999 through 1/1/2012. The ZigZag indicator connects highs and lows of a specified minimum change, ignoring smaller intermediate changes. See Figure 6.1 for an example of a 3% zigzag applied to closing prices. The minimum percentage change from a bottom to top or a top to bottom is at least 3%, and all changes between extremes are less than 3%.

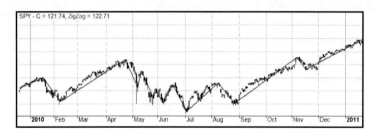

Figure 6.1 -- 3% ZigZag

An ideal entry would be to buy at the close on the day that forms the zigzag bottom, and an ideal exit to sell at the close of the day that forms the zigzag top. Without knowledge of the future, that is impossible. But it is useful for research purposes to take nearly ideal entries using forward knowledge (an intentional future leak), then test and compare a variety of exit techniques.

When testing trading systems, I ordinarily set commission and slippage to zero, then adjust after the logic has been designed and when tactics for trading are being decided. For this study it is possible that the zigzag percentage will be small and average trades similarly small. To avoid completely unrealistic results, commission is set to $5 per trade—that is $5 to enter and $5 to exit. For trades made as Market on Close, slippage is set to zero.

Any system that buys the zigzag bottom is taking a mean reversion entry—clearly the price at the bottom is less than any moving average whose length is greater than 1.0. Subsequent prices will be increasing and reverting toward the mean. But people trading systems based on trend following, seasonality, cycles, or patterns would be just as pleased if they could buy at or near the ideal bottom, so this study helps identify the practical exit techniques for any system.

Being late or early with the entry makes a difference. The best entries are made on the exact zigzag bottom. In the sections that follow, entry to long positions will be made at the ideal bottom, as well as a few days early and late. Accuracy of picking the bottom is most important when the amount of the trade being captured is small—less than about 2%. As the potential trade size increases, accuracy of entry is less important.

Exits can be described as belonging to one of several categories. Each will be studied in turn. All discussions, examples, and program code are in terms of long positions, but everything applies equally well to short positions. In all cases, the entry is assumed to have been made at the close of the signal bar at the closing price.

LOGIC

In many, perhaps most, systems, the primary technique to exit a trade is a logical statement or a rule, such as exit when the price rises enough to cross its moving average, or when an indicator reaches a critical level.

MOVING AVERAGE

Just for this one technique, there are many choices including:
- Base the calculations on the close or on intra-day price.
- Use a simple moving average, exponential, weighted, adaptive, butterworth, etc.
- Select the length of the moving average lookback period.
- Exit intra-day or at the close.

Because you will want to do your own research using your own entry

Exits

method and your own trading settings, the results of only a few options are discussed in detail. AmiBroker programs are provided to illustrate how the technique is implemented.

Listing 6.1 shows the AmiBroker code that allows easy testing and comparison of exits based on moving averages, including simple, exponential, and adaptive. A check is made before entry to be certain the price is below the exit moving average. Since both the daily price and the moving average change as additional bars are added to the price series, the cross of the price up through its moving average will definitely happen, and no other exits are used.

```
//   ZigZagBottomsWithMAExits.afl
//
//   Enter a few days before or after a zigzag bottom.
//   Test alternative exits based on moving averages.
//

SetOption( "ExtraColumnsLocation", 1 );
SetOption ( "CommissionMode", 2 );  // $ per trade
SetOption( "CommissionAmount", 5 );
SetOption( "InitialEquity", 100000 );
SetPositionSize( 10000, spsValue );
MaxPos = 1;
SetOption( "MaxOpenPositions", MaxPos );
SetTradeDelays( 0, 0, 0, 0 );
BuyPrice = Close;
SellPrice = Close;
//   ObFn == K-ratio, CAR/MDD, expectancy

ZZPercent = 1.2;  //Optimize("ZZPercent",1.2,1,11,1);
ZZ = Zig( C, ZZPercent );
ZZBottom = ( ZZ < Ref( ZZ, -1 ) ) AND ( ZZ < Ref( ZZ, 1 ) );

NumberBottoms = Sum( ZZBottom, 252 );

//   EntryOffset is the number of days
//   between entry and the ZigZag bottom.
//   A negative value means the bottom was
//   in the past -- the entry is late.
EntryOffset = Optimize( "EntryOffset", 0, -3, 3, 1 );

////////////  Exits  ////////////////

//   The length of the lookback for the average
ExitMALength = Optimize( "ExitMALength", 2, 2, 10, 1 );

//   The type of average
ExitMethod = Optimize( "ExitMethod", 1, 1, 3, 1 );

switch ( ExitMethod )
{
```

```
case 1:
//  Simple moving average
    ExitMA = MA( C, ExitMALength );
    Sell = Cross( C, ExitMA );
    break;

case 2:
//  Exponential moving average
    ExitMA = EMA( C, ExitMALength );
    Sell = Cross( C, ExitMA );
    break;

case 3:
//  Adaptive moving average
    ExitMA = AMA( C, ExitMALength );
    Sell = Cross( C, ExitMA );
    break;

default:
    ExitMA = MA( C, ExitMALength );
    Sell = Cross( C, ExitMA );
    break;
}

Buy = Ref( ZZBottom, EntryOffset ) AND C < ExitMA;

////////  Plots //////////

Plot( C, "C", colorBlack, styleCandle );
Plot(ExitMA,"ExitMA",colorGreen,styleLine);
Plot( ZZ, "ZigZag", colorBlue, styleLine | styleThick );

shapes = IIf( Buy, shapeUpArrow, IIf( Sell, shapeDownArrow,shapeNone));
shapecolors = IIf( Buy, colorGreen, IIf( Sell, colorRed, colorWhite));
PlotShapes( shapes, shapecolors );

Plot( NumberBottoms, "NumberBottoms", colorRed, styleLine |
        styleOwnScale, 0, 50 );
Plot( 25, "", colorRed, styleLine | styleOwnScale, 0, 50 );

////////////  end ///////////////
```

Listing 6.1 -- Test Moving Average Exit

When entries are based on the zigzag bottoms of various percentages, with a range of entry accuracy from 3 days early to 3 days late, the best single moving average exit is a simple moving average with a lookback length between 5 and 8 days. Use longer lookback when the entry is early or late, shorter when it is accurate. Exponential moving average worked about as well as simple. Adaptive moving average performed poorly.

You can, and should, do your own research, using your own entry logic.

Exits

RSI INDICATOR

Exit could be made based on indicators other than moving average—perhaps an indicator such as the RSI rising through a level. Again, there are many alternatives:
- Which indicator.
- What lookback length or other parameter value.
- What indicator level.
- What transformations are applied.
- Is the level static or dynamic.
- Exit at the close or intra-day.

Listing 6.2 shows the AmiBroker code that can be used as the basis for testing exits based on the RSI indicator.

```
//    ZigZagBottomsWithRSIExits.afl
//
//    Enter a few days before or after a zigzag bottom.
//    Test alternative exits based on RSI levels.
//

SetOption( "ExtraColumnsLocation", 1 );
SetOption ( "CommissionMode", 2 );  // $ per trade
SetOption( "CommissionAmount", 5 );
SetOption( "InitialEquity", 100000 );
SetPositionSize( 10000, spsValue );
MaxPos = 1;
SetOption( "MaxOpenPositions", MaxPos );
SetTradeDelays( 0, 0, 0, 0 );
BuyPrice = Close;
SellPrice = Close;
//    ObFn == K-ratio, CAR/MDD, expectancy

ZZPercent = 1.2; //Optimize("ZZPercent",1.2,1,11,1);
ZZ = Zig( C, ZZPercent );
ZZBottom = ( ZZ < Ref( ZZ, -1 ) ) AND ( ZZ < Ref( ZZ, 1 ) );

NumberBottoms = Sum( ZZBottom, 252 );

//    EntryOffset is the number of days
//    between entry and the ZigZag bottom.
//    A negative value means the bottom was
//    in the past -- the entry is late.
EntryOffset = Optimize( "EntryOffset", 0, -3, 3, 1 );

////////////    Exits    ////////////////

//    The length of the lookback for the RSI
ExitRSILookback = Optimize( "ExitRSILookback", 2, 2, 8, 1 );
RSIValue = RSIa( C, ExitRSILookback );

//    The RSI exit level
ExitRSILevel = Optimize( "ExitRSILevel", 50, 0, 100, 1 );
```

```
Buy = Ref( ZZBottom, EntryOffset ) AND RSIValue < ExitRSILevel;
Sell = Cross( RSIValue, ExitRSILevel );

////////   Plots   //////////

Plot( C, "C", colorBlack, styleCandle );
Plot( ZZ, "ZigZag", colorBlue, styleLine | styleThick );
Plot( RSIValue, "RSIValue", colorGreen, styleLine | styleOwnScale );

shapes = IIf( Buy, shapeUpArrow, IIf( Sell, shapeDownArrow,shapeNone));
shapecolors = IIf( Buy, colorGreen, IIf( Sell, colorRed, colorWhite));
PlotShapes( shapes, shapecolors );

Plot( NumberBottoms, "NumberBottoms", colorRed, styleLine |
      styleOwnScale, 0, 50 );
Plot( 25, "", colorRed, styleLine | styleOwnScale, 0, 50 );

/////////////   end   ////////////////
```

Listing 6.2 -- Test RSI Indicator Exit

If you are 3 days early, RSI exit does not work well. From 2 days early to 3 days late, a generally good exit is signaled when the 2 period RSI crosses up through 65 to 75, or when the 3 period RSI crosses up through about 60.

Z-SCORE INDICATOR

The z-score of a data point is the distance that point is relative to the mean of a set of similar data points. The unit of measure is the standard deviation of the set of data. Listing 6.3 shows the AmiBroker code that can be used as the basis for testing exits based on the z-score indicator.

```
//   ZigZagBottomsWithZScoreExits.afl
//
//   Enter a few days before or after a zigzag bottom.
//   Test alternative exits based on Z Score levels.
//

SetOption( "ExtraColumnsLocation", 1 );
SetOption ( "CommissionMode", 2 );  // $ per trade
SetOption( "CommissionAmount", 5 );
SetOption( "InitialEquity", 100000 );
SetPositionSize( 10000, spsValue );
MaxPos = 1;
SetOption( "MaxOpenPositions", MaxPos );
SetTradeDelays( 0, 0, 0, 0 );
BuyPrice = Close;
SellPrice = Close;
//   ObFn == K-ratio, CAR/MDD, expectancy

ZZPercent = 5.0; //Optimize("ZZPercent",1.2,1,11,1);
ZZ = Zig( C, ZZPercent );
ZZBottom = ( ZZ < Ref( ZZ, -1 ) ) AND ( ZZ < Ref( ZZ, 1 ) );

NumberBottoms = Sum( ZZBottom, 252 );
```

```
//    EntryOffset is the number of days
//    between entry and the ZigZag bottom.
//    A negative value means the bottom was
//    in the past -- the entry is late.
EntryOffset = Optimize( "EntryOffset", 0, -3, 3, 1 );

////////////   Exits   //////////////

//  The length of the lookback for the Z Score
ExitZScoreLookback = Optimize( "ExitZScoreLookback", 2, 2, 20, 1 );
ZScore = ( C - MA( C, ExitZScoreLookback ) ) /
           StDev( C, ExitZScoreLookback );

//   z-score level
ExitZScoreLevel = 0.0; //Optimize( "ExitZScoreLevel", 0, -1, 2, 0.1 );

Buy = Ref( ZZBottom, EntryOffset ) AND ZScore < ExitZScoreLevel;
Sell = Cross( ZScore, ExitZScoreLevel  );

/////////   Plots  ///////////

Plot( C, "C", colorBlack, styleCandle );
Plot( ZScore, "zscore", colorGreen, styleLine | styleOwnScale );
Plot( ZZ, "ZigZag", colorBlue, styleLine | styleThick );

shapes = IIf( Buy, shapeUpArrow, IIf( Sell, shapeDownArrow,shapeNone));
shapecolors = IIf( Buy, colorGreen, IIf( Sell, colorRed, colorWhite));
PlotShapes( shapes, shapecolors );

Plot( NumberBottoms, "NumberBottoms", colorRed, styleLine |
        styleOwnScale, 0, 50 );
Plot( 25, "", colorRed, styleLine | styleOwnScale, 0, 50 );

////////////   end   ////////////////
```

Listing 6.3 -- Test Z-Score Indicator Exit

Since z-score of 0.0 represents the mean, and the systems we are testing are mean reversion systems, exiting when the z-score equals 0.0 might be a good place to begin. For the entire range of entry accuracy, from 3 days early to 3 days late, a lookback length of 8 or 9 with z-score crossing upward through 0.0 works well.

Relaxing the 0.0 restriction and testing all combinations of lookback length with levels from -1.0 to +2.0, a lookback of 8 or 9 with a level of about 1.0 works as well or better than crossing 0.0. The interpretation is to hold the long position until the closing price has not only returned to the mean, but has risen above it by 1.0 standard deviations.

SIMILARITY OF INDICATORS

For those indicators that use explicit lookback lengths, using short lookback periods (typically 2 through 10 for mean reversion systems) causes

indicator values to change quickly. Through selection of the indicator, its length, and the action level, any one of the commonly used indicators is adequate. Pick your favorite from among:
- RSI.
- CCI.
- Stochastic.
- William's R.
- Percent B.
- Z-score.

The possibilities are endless. As a start, use one of the logic methods tested here, or add your favorite. If possible, select a method and its parameters that are robust enough that neither the logic nor its parameters need to be included among those searched during the synchronization process as the entry methods are selected. Three possibilities, based on the studies discussed so far, are:
- Close crossing up through the MA(C,4).
- 2 period RSI crossing up through 70.
- 9 day z-score crossing up through 1.0.

Although any of these methods work well, it is not my recommendation that a combination of all of them be used. As with any component of a trading system, the specific exit rule used in a system must come into play often enough for the statistics associated with its use to be meaningful.

Holding Period

Systems that often reach their goals in a short period of time sometimes benefit from an exit based on holding period. Assuming a long trade has been entered at the close of a day of trading, some of the choices for the timed exit are:
- Next Day's Open (NDO) or First open—the open of trading following the close at which the trade was entered. This would be a single overnight session. When an accurate system gives an accurate signal, the gain during the first overnight period is often significant. (See *Modeling Trading System Performance* for a more detailed discussion of the positive value of holding overnight.)
- First close.
- Second open.
- Second close.
- nth open.
- nth close.

- First profitable open.
- First profitable close.

Listing 6.4 shows the AmiBroker code to exit after an n-day holding period. Note the variable named "useClose" which takes values of 1 or 0. When set to 1, the exit will be at the Close on the nth day following the entry; when set to 0, the exit will be at the Open on the nth day following the entry. There is no exit logic—no "sell rule." The Apply-Stop statement implements the holding period exit.

```
//   ZigZagBottomsWithNDayHold.afl
//
//   Enter a few days before or after a zigzag bottom.
//   Test exit after an n day holding period.
//

SetOption( "ExtraColumnsLocation", 1 );
SetOption ( "CommissionMode", 2 ); // $ per trade
SetOption( "CommissionAmount", 5 );
SetOption( "InitialEquity", 100000 );
SetPositionSize( 10000, spsValue );
MaxPos = 1;
SetOption( "MaxOpenPositions", MaxPos );
SetTradeDelays( 0, 0, 0, 0 );
BuyPrice = Close;
//SellPrice = Close;
//   ObFn == K-ratio, CAR/MDD, expectancy

ZZPercent = 1.2; //Optimize("ZZPercent",1.2,1,11,1);
ZZ = Zig( C, ZZPercent );
ZZBottom = ( ZZ < Ref( ZZ, -1 ) ) AND ( ZZ < Ref( ZZ, 1 ) );

NumberBottoms = Sum( ZZBottom, 252 );

//   EntryOffset is the number of days
//   between entry and the ZigZag bottom.
//   A negative value means the bottom was
//   in the past -- the entry is late.
EntryOffset = 0; //Optimize( "EntryOffset", 0, -3, 3, 1 );

////////////   Exits   ////////////////

Buy = Ref( ZZBottom, EntryOffset );
Sell = 0;

useClose = 0; //Optimize("useClose", 1,0,1,1);

if ( useClose == 1 )
{
    SellPrice = Close;
}
else
{
    SellPrice = Open;
}
```

```
//   Set holding period
HoldDays = 2; //Optimize("HoldDays", 2, 1, 20, 1);

ApplyStop( stopTypeNBar, stopModeBars, HoldDays );

////////   Plots   //////////

Plot( C, "C", colorBlack, styleCandle );
Plot( ZZ, "ZigZag", colorBlue, styleLine | styleThick );

shapes = IIf( Buy, shapeUpArrow, IIf( Sell, shapeDownArrow,shapeNone));
shapecolors = IIf( Buy, colorGreen, IIf( Sell, colorRed, colorWhite));
PlotShapes( shapes, shapecolors );

Plot( NumberBottoms, "NumberBottoms", colorRed, styleLine |
        styleOwnScale, 0, 50 );
Plot( 25, "", colorRed, styleLine | styleOwnScale, 0, 50 );

/////////////   end   ////////////////
```

Listing 6.4 -- Test N-Day Holding Period Exit

Listing 6.5 shows the program that exits at the first profitable open.

```
//   ZigZagBottomsWithFirstProfitableOpen.afl
//
//   Enter a few days before or after a zigzag bottom.
//   Test exit at the first profitable open.
//

SetOption( "ExtraColumnsLocation", 1 );
SetOption ( "CommissionMode", 2 ); // $ per trade
SetOption( "CommissionAmount", 5 );
SetOption( "InitialEquity", 100000 );
SetPositionSize( 10000, spsValue );
MaxPos = 1;
SetOption( "MaxOpenPositions", MaxPos );
SetTradeDelays( 0, 0, 0, 0 );
BuyPrice = Close;
//SellPrice = Close;
//   ObFn == K-ratio, CAR/MDD, expectancy

ZZPercent = 1.2; //Optimize("ZZPercent",1.2,1,11,1);
ZZ = Zig( C, ZZPercent );
ZZBottom = ( ZZ < Ref( ZZ, -1 ) ) AND ( ZZ < Ref( ZZ, 1 ) );

NumberBottoms = Sum( ZZBottom, 252 );

//   EntryOffset is the number of days
//   between entry and the ZigZag bottom.
//   A negative value means the bottom was
//   in the past -- the entry is late.
EntryOffset = Optimize( "EntryOffset", 0, -3, 3, 1 );

////////////   Exits   //////////////

Buy = Ref( ZZBottom, EntryOffset );
```

```
EntryPrice = ValueWhen( Buy, BuyPrice );
ProfitableOpen = Open > EntryPrice;

Sell = ProfitableOpen;
SellPrice = Open;

////////   Plots  ///////////

Plot( C, "C", colorBlack, styleCandle );
Plot( ZZ, "ZigZag", colorBlue, styleLine | styleThick );

shapes = IIf( Buy, shapeUpArrow, IIf( Sell, shapeDownArrow,shapeNone));
shapecolors = IIf( Buy, colorGreen, IIf( Sell, colorRed, colorWhite));
PlotShapes( shapes, shapecolors );

Plot( NumberBottoms, "NumberBottoms", colorRed, styleLine |
        styleOwnScale, 0, 50 );
Plot( 25, "", colorRed, styleLine | styleOwnScale, 0, 50 );

//////////////   end  ////////////////
```

Listing 6.5 -- Test First Profitable Open Exit

Since this exit insists on a profit, all trades are winners, even those entered early that were unprofitable with some of the other exits.

Profit Target

A profit target can be set at some percentage or price above the entry price. When the trading price reaches the target price, exit from the trade:
- At that price using an intra-day limit order.
- At the close when the close is greater than or equal to the target price.
- At the close when the intra-day high is greater than or equal to the target price.

As the target is decreased from a level that is never hit down to the minimum change that is tradable, the percentage of trades that are exited at the target price increases. There is a trade off—tighter targets produce higher percentage winning trades, but lower percentage won on each trade. At the tightest level, the only trades that are losers (before commission and slippage) are those that gap down in the first overnight and never recover. See the section Viewing Trades as Bars later in this chapter for an illustration of how use of a profit target affects trades.

When the target is either very loose and seldom hit, or very tight and nearly always hit, very little total profit is realized. Plotting the total profit versus target percentage, the graph looks like the St. Louis arch—zero on each end and high in the middle. The optimum target level is somewhere between the two extremes. If the profit target is greater than the potential profit (which is the gain of the zigzag percentage in the test program), the target is never hit and using a profit target is not useful as an exit technique in this case. If the profit target is less than the zigzag percentage, many trades are closed at a relatively low profit. There is a nearly linear relationship between the percentage of profit captured and the percentage that the profit target is of the zigzag amount. For example, if the zigzag percent is 5% and the profit target is set at 2%, only about 40% of the potential profit is captured.

If you are looking for systems that have a high percentage of winning trades, consider using a combination of a profit target that gives you a satisfactory profit after covering your commission and slippage, along with patterns that lead to trades that, if allowed to exit at their maximum favorable excursion, MFE, would have profits about two times the profit target. For example, a profit target of 0.5% gives a gross profit of $50 for a $10,000 trade, netting $40 after commission. Look for patterns that leads to 1% trades. For SPY, there are about 20 to 30 long trades that gain 1.2% or more in a typical year.

In general:
- If used at all, the profit target exit must be the reason for the exit enough times to be testable on its own.
- Use profit targets when the potential trade is relatively small—less than about 2%.
- Be as accurate on entry day as possible—the profit from hitting the exact bottom returns four to ten times the profit from being one day early or late, and ten or more times the profit of being two days early or late.
- Size the profit target to the magnitude of the trade, trying to exit before the point of highest profit.
- Be aware of intra-trade drawdown. Intra-trade drawdown increases as trades are held longer or as higher percentage gains are sought. For example, a 5% profit target based on 5% zigzag allows an intra-trade drawdown of 4.9%. When based on actual SPY prices, there is no zigzag bottom to indicate the trade is over, and the size of the intra-trade drawdown can exceed the profit target.

Listing 6.6 shows the AmiBroker program to test profit targets set as percentages.

```
//    ZigZagBottomsWithProfitTarget.afl
//
//    Enter a few days before or after a zigzag bottom.
//    Test profit target exits.
//

SetOption( "ExtraColumnsLocation", 1 );
SetOption ( "CommissionMode", 2 );  // $ per trade
SetOption( "CommissionAmount", 5 );
SetOption( "InitialEquity", 100000 );
SetPositionSize( 10000, spsValue );
MaxPos = 1;
SetOption( "MaxOpenPositions", MaxPos );
SetTradeDelays( 0, 0, 0, 0 );
BuyPrice = Close;
SellPrice = Close;
//    ObFn == K-ratio, CAR/MDD, expectancy

ZZPercent = 1.2; //Optimize("ZZPercent",1.2,1,11,1);
ZZ = Zig( C, ZZPercent );
ZZBottom = ( ZZ < Ref( ZZ, -1 ) ) AND ( ZZ < Ref( ZZ, 1 ) );

NumberBottoms = Sum( ZZBottom, 252 );

//    EntryOffset is the number of days
//    between entry and the ZigZag bottom.
//    A negative value means the bottom was
//    in the past -- the entry is late.
EntryOffset = Optimize( "EntryOffset", 0, -3, 3, 1 );

///////////   Exits   ///////////////////

Buy = Ref( ZZBottom, EntryOffset );
Sell = 0;

// Profit target in percentage

ProfitTarget = Optimize( "ProfitTarget", 1.0, 0.5, 11.0, 0.5 );
ApplyStop( stopTypeProfit, stopModePercent, ProfitTarget );

////////   Plots   ///////////

Plot( C, "C", colorBlack, styleCandle );
//Plot(ExitMA,"ExitMA",colorGreen,styleLine);
Plot( ZZ, "ZigZag", colorBlue, styleLine | styleThick );

shapes = IIf( Buy, shapeUpArrow, IIf( Sell, shapeDownArrow,shapeNone));
shapecolors = IIf( Buy, colorGreen, IIf( Sell, colorRed, colorWhite));
PlotShapes( shapes, shapecolors );

Plot( NumberBottoms, "NumberBottoms", colorRed, styleLine |
       styleOwnScale, 0, 50 );
Plot( 25, "", colorRed, styleLine | styleOwnScale, 0, 50 );

//////////////   end   ///////////////
```

Listing 6.6 -- Test Profit Target Exit

Figure 6.2 shows the relationship between trades that have a potential of 1% to 2%, produced by using a 1% zigzag, and exits using a profit target ranging from 0.5% to 5%. Note the peak at a profit target of 1.0 to 1.5%. In general, it is a good idea to set the profit target level close to, but lower than, the profit expected to be available from the trades. If the profit target is set higher it will seldom be hit, and consequently it will be an ineffective exit technique. Also note the importance of accuracy of entry. The high profits are associated with entry at the bottom.

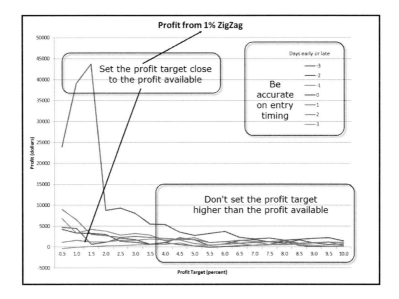

Figure 6.2 -- Profit target related to ZigZag of 1%

Figure 6.3 shows the equity curve that results when a 1% profit target is used with trades that typically have a profit of at least 1% available.

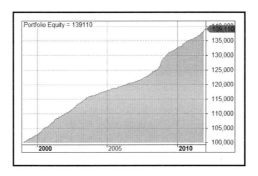

Figure 6.3 -- Equity curve 1% profit target and 1% zigzag

Figure 6.4 shows the relationship between trades that have a potential of at least 5%, produced by a 5% zigzag, and exits using a profit target ranging from 0.5% to 5%. Note that accuracy is still important. But less so than when the potential trade is smaller.

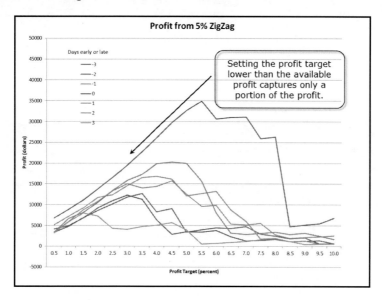

Figure 6.4 -- Profit target related to ZigZag of 5%

Figure 6.5 shows the equity curve when a 1% profit target is used with trades that typically have a 5% profit potential. Compared with Figure 9, total profit is lower (8,800 versus 39,100) and there are longer periods of flat equity.

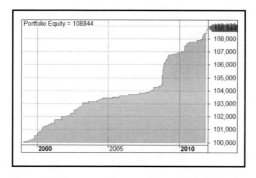

Figure 6.5 -- Equity curve 1% profit target and 5% zigzag

DYNAMIC PROFIT TARGETS

The profit targets described above use static targets — a given percentage without regard for the volatility of the price. Fluctuations in volatility might be useful in setting profit targets that adapt to changing conditions. A sliding window of recent prices is used to compute the volatility— measured in this example by the average true range. The profit target is set in proportion to the ATR. Listing 6.7 shows the AmiBroker code to test dynamic profit targets based on ATR.

```
//    ZigZagBottomsWithATRProfitTargets.afl
//
//    Enter a few days before or after a zigzag bottom.
//    Test profit target using adaptive ATR range.
//

SetOption( "ExtraColumnsLocation", 1 );
SetOption ( "CommissionMode", 2 ); // $ per trade
SetOption( "CommissionAmount", 5 );
SetOption( "InitialEquity", 100000 );
SetPositionSize( 10000, spsValue );
MaxPos = 1;
SetOption( "MaxOpenPositions", MaxPos );
SetTradeDelays( 0, 0, 0, 0 );
BuyPrice = Close;
SellPrice = Close;
//    ObFn == K-ratio, CAR/MDD, expectancy

ZZPercent = 5.0; //Optimize("ZZPercent",1.2,1,11,1);
ZZ = Zig( C, ZZPercent );
ZZBottom = ( ZZ < Ref( ZZ, -1 ) ) AND ( ZZ < Ref( ZZ, 1 ) );

NumberBottoms = Sum( ZZBottom, 252 );

//    EntryOffset is the number of days
//    between entry and the ZigZag bottom.
//    A negative value means the bottom was
//    in the past -- the entry is late.
EntryOffset = Optimize( "EntryOffset", 0, -3, 3, 1 );

///////////// Exits /////////////////

Buy = Ref( ZZBottom, EntryOffset );
Sell = 0;

//    Set profittarget based on recent volatility
ATRMult = Optimize( "ATRMult", 1.0, 0.2, 3.0, 0.2 );
ATRLookback = Optimize( "ATRLookback", 5, 1, 20, 1 );
ProfitTarget = ATRMult * ATR( ATRLookback );
ApplyStop( stopTypeProfit, stopModePoint, ProfitTarget );

////////  Plots ///////////

Plot( C, "C", colorBlack, styleCandle );
Plot( ZZ, "ZigZag", colorBlue, styleLine | styleThick );

shapes = IIf( Buy, shapeUpArrow, IIf( Sell, shapeDownArrow,shapeNone));
shapecolors = IIf( Buy, colorGreen, IIf( Sell, colorRed, colorWhite));
PlotShapes( shapes, shapecolors );
```

```
Plot( NumberBottoms, "NumberBottoms", colorRed, styleLine |
    styleOwnScale, 0, 50 );
Plot( 25, "", colorRed, styleLine | styleOwnScale, 0, 50 );

////////////    end    ////////////////
```

Listing 6.7 -- Test Dynamically Computed Profit

It would be nice to report that using a dynamic profit target of x times the n day ATR worked well in most cases. But I did not find values for n, the lookback period, or x, the multiplier, that were generally good parameters. What I did find was that when the profit potential is small, x and n are relatively small; when the profit potential is greater, x and n are greater. Even at that, there was wide variation with x ranging from 0.4 to 2.8 and n ranging from 1 to 19. If you plan to use a dynamic profit target based on ATR, test carefully to be certain the values you use are robust, and not simply curve-fit to the data.

TRAILING EXIT

The trailing exit is sometimes called a trailing stop. While it is true that the order used for execution is a stop order, I prefer to think of the exit technique. To be effective, the initial level of a trailing exit must be far enough below the entry that it will not be hit in the first day or two. It is intended to be a trailing exit, not a maximum loss exit, although it will act as one if price drops early in the trade. As prices rise, the trailing exit price rises; as prices hesitate, remain flat, or decline, the trailing exit does not decline. Some implementations, such as chandelier, remain flat, continuing upward only when the price continues upward. Other implementations, such as parabolic, continue rising even when the price is declining. When the low of a bar declines enough to be at or below the level of the trailing exit, the trade is closed at the exit price. Unless there is also some other exit technique, exits made using a trailing exit with the chandelier method will always give back some open profit before the exit. Since the parabolic exit point continues to move, it might rise enough to meet a stalled trading price and exit with little or no give back of open profit.

When daily bars are used, swing trades that are held only a few days often do not have enough time and price range to both establish the exit level and have the exit level follow the price. If multiple time frames are used, there is much more flexibility, and a trailing exit can be implemented on a finer time scale, say using hourly bars.

Listing 6.8 shows the program to exit using a trailing exit. Note that there are two methods illustrated in the program — one uses a number of points, the other a percentage.

```
//    ZigZagBottomsWithTrailingExit.afl
//
//    Enter a few days before or after a zigzag bottom.
//    Use a trailing exit.
//

SetOption( "ExtraColumnsLocation", 1 );
SetOption ( "CommissionMode", 2 ); // $ per trade
SetOption( "CommissionAmount", 5 );
SetOption( "InitialEquity", 100000 );
SetPositionSize( 10000, spsValue );
MaxPos = 1;
SetOption( "MaxOpenPositions", MaxPos );
SetTradeDelays( 0, 0, 0, 0 );
BuyPrice = Close;
SellPrice = Close;
//    ObFn == K-ratio, CAR/MDD, expectancy

ZZPercent = 1.2; //Optimize("ZZPercent",1.2,1,11,1);
ZZ = Zig( C, ZZPercent );
ZZBottom = ( ZZ < Ref( ZZ, -1 ) ) AND ( ZZ < Ref( ZZ, 1 ) );

NumberBottoms = Sum( ZZBottom, 252 );

//    EntryOffset is the number of days
//    between entry and the ZigZag bottom.
//    A negative value means the bottom was
//    in the past -- the entry is late.
EntryOffset = 0; //Optimize( "EntryOffset", 0, -3, 3, 1 );

////////////    Exits    ////////////////

Buy = Ref( ZZBottom, EntryOffset );
Sell = 0;

//    Two methods of setting the trailing exit.
//    Use either
//    Or use both
//    Each will be adjusted bar-by-bar.
//    Whichever is closer will be used.

//    Set the trail amount in points based on recent volatility
ATRMult = Optimize( "ATRMult", 1.2, 0.2, 3.0, 0.2 );
ATRLookback = Optimize( "ATRLookback", 6, 1, 20, 1 );
TrailPoints = ATRMult * ATR( ATRLookback );
ApplyStop( stopTypeTrailing, stopModePoint, TrailPoints );

//    Set the trail amount as a fixed percentage
TrailPercent = 1.0;
ApplyStop( stopTypeTrailing, stopModePercent, TrailPercent );

////////    Plots    ///////////
```

Exits

```
Plot( C, "C", colorBlack, styleCandle );
Plot( ZZ, "ZigZag", colorBlue, styleLine | styleThick );

shapes = IIf( Buy, shapeUpArrow, IIf( Sell, shapeDownArrow,shapeNone));
shapecolors = IIf( Buy, colorGreen, IIf( Sell, colorRed, colorWhite));
PlotShapes( shapes, shapecolors );

Plot( NumberBottoms, "NumberBottoms", colorRed, styleLine |
        styleOwnScale, 0, 50 );
Plot( 25, "", colorRed, styleLine | styleOwnScale, 0, 50 );

///////////// end /////////////////
```

Listing 6.8 -- Test Trailing Exit

Try using a small multiplier, say 1.5, and a short lookback, say 6 days, with the ATR calculation for swing trading systems. Or a small percentage, say 1.0%, for the fixed percentage version.

MAXIMUM LOSS EXIT

Maximum loss exits hurt swing trading systems. They usually hurt trend following systems as well. To see the effect they have on your system, code the system without a maximum loss exit and note the system performance. Add a maximum loss exit that is so far below the entry that it is never hit. Incrementally tighten the exit so that it is the cause of an increasing number of trades and note the changes in system performance.

When the maximum loss level is close to the entry, every trade whose maximum adverse excursion (MAE) touches the exit level will be a losing trade, even if it would later have recovered to profitability.

VIEWING TRADES AS BARS

Just as a price bar is created by combining the open, high, low, and closing prices of a period, a trade bar can be created by combining the account equity at the open of the trade, high intra-trade equity, low intra-trade equity, and closing equity. Trade bars can be displayed in either candlestick format or barchart format. Figure 6.6 shows an illustration of converting account equity into a trade bar.

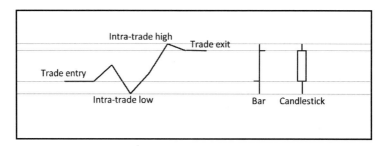

Figure 6.6 -- Forming a trade bar from the account equity

Profit targets and maximum loss exits both act to truncate trade bars. Figure 6.7 shows some trades where neither profit target nor maximum loss exit was used. The bars are aligned by the entry. The upper and lower wicks are the intra-trade high equity and low equity. Each trade would have closed at the edge of the body above or below the trade entry line in the absence of either a profit target or maximum loss exit. The light horizontal lines represent the profit target and maximum loss stop discussed in the next sections.

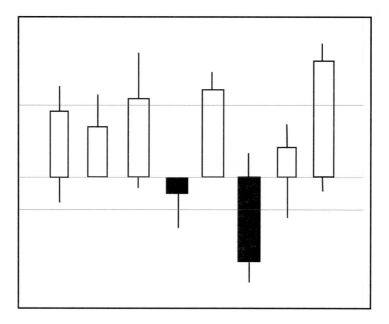

Figure 6.7 -- Raw trades

A profit target exit ensures that the profit at the target level is realized whenever intra-trade profit reaches the target level, but prevents taking any additional profit. Since the profit target is implemented using a limit order, the fill will be at the target price. Figure 6.8 illustrates how trade equity is truncated by using profit targets. The lower wick has been removed since there is no way to know the sequence of intra-trade prices. Any body or wick that extended above the profit target has been removed because the trade was closed at the profit target before that higher price could be reached. One condition that allows a trade to be closed with a profit higher than the target is a gap opening above the target level.

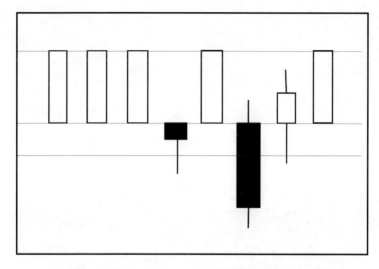

Figure 6.8 -- Trades are truncated by use of profit targets

A maximum loss exit attempts to limit maximum loss by placing a loss limit below the entry price. Every trade that experiences an intra-trade loss of at least that amount will be closed at a loss. Since the maximum loss exit is implemented using a stop order, the realized loss might be greater due to slippage on the execution. Figure 6.9 illustrates how trades are truncated by using maximum loss exits, and how trades that would have closed at a profit can become losers.

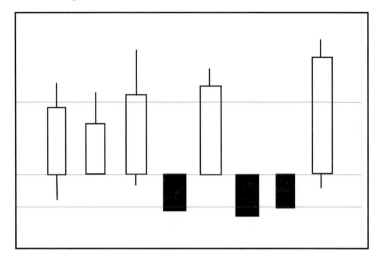

Figure 6.9 -- Trades are truncated by use of maximum loss exits

SUMMARY

As I explain in *Modeling Trading System Performance*, the sweet spot for a trading system has these characteristics:
- Trading frequently. 20 to 30 times a year.
- Holding a short period. 1 to 3 days.
- Winning a high percentage of trades. At least 65%
- Limiting the size of losses on losing trades.

Given reasonable entries, the first three, and to some extent the fourth, can be achieved by including one or more of four exit techniques in every trading system:
- Logic. Rules, such as exit at the cross of price with a moving average, or when the value of an indicator reaches a significant level, or recognition of a pattern associated with good exits.
- Profit target. Set so that it captures a reasonable portion of the profit potential of each trade, provides enough profit to cover commission and slippage, and is hit often.
- Holding period. Set so that it exits trades that are unlikely to increase profit given longer holding.
- Trailing exit. Set to protect a substantial portion of maximum intra-trade profit.

Based on the test programs listed above, and on tests run on many systems, any of the logic exit rules discussed, or alternatives of your choice, is reasonable. Adding a profit target, a maximum holding period, and / or a trailing exit to the logic exits may or may not improve the system. It is reasonable to design systems that have no logic exit—only profit target, holding period, and / or trailing exit.

When setting the logic and parameters for exits, determine how they are best configured for your system, and fix as many of the parameters as you can. Avoid adding superfluous exit logic and complexity to the system.

Limiting losses on losing trades is the more difficult challenge. Some techniques that help include:
- Trailing exit. Try a trailing exit even if your system uses daily bars and trades only at the close. Being able to exit intra-day will improve performance, even when using daily bars. Use of intra-day bars will help significantly.
- Avoid conditions that have high risk. Improve entries. Use filters to identify favorable and unfavorable conditions.
- Position size. Analyze the system carefully, and set your position size so that you can afford the losses that occur in the tail of the distribution.
- Options. Rather than trading the underlying, use debit options to take the positions.

Do not use maximum loss stops with mean reversion / swing trade systems. The advice to know your maximum risk exit price and place a stop order at that price when you enter your position is bad advice.

If the trade cannot be entered at the exact bottom, it is usually better to be a day late than a day early—particularly try not to be more than two days early for swing trades.

Analyze the effect of taking additional positions. If you are in a long trade and the system gives another buy signal, the new signal is probably of higher quality than the first and you should probably take it. (Do the research and analysis to verify this for your specific system.) I do not consider this to be either scaling-in or a portfolio. I consider this to a separate system which should be analyzed on its own merits. There are special considerations when trading multiple systems, particularly when the trades are strongly correlated. Read Chapter 8 discussing the considerations of systems of systems.

7

Entries

Entries are relatively easy—most of the work has been done computing the indicators and identifying the patterns that precede a profitable trade, signaling a Buy. The entry to a long positions is one of these:
- Act at the close of the bar / day that generates the signal, at the closing price of that bar. MOC Order (MOC of the action bar).
- Act at the open of the next day, at the opening price. NDO, or MOO Order.
- Place a limit order, good for some period of time or number of bars, to enter a position at a price of your choosing or better. Limit Order.
- Place a stop order, good for some period of time or number of bars, to enter a long position at the market price if the issue trades at that price or higher. Stop Order.

Any of the first three are commonly used in mean reversion systems. The fourth is commonly used in trend following systems, and can be useful in mean reversion system when the setup uses longer length bars, say daily, and the entry uses shorter length bars, say hourly.

MARKET ON CLOSE OF ACTION BAR

My research shows that the signals from a good system should be acted upon as soon as possible. If a system reliably forecasts the direction of price from the close of one day to the close of the next day, about one-third of that price change happens in the overnight market. Entering MOC of the action bar captures that profit—entering NDO misses it.

The AmiBroker code to use is:

```
//   Enter MOC of action bar
SetTradeDelays( 0, 0, 0, 0 );
BuyPrice = Close;
SellPrice = Close;
ER1 = DayOfWeek() == 2;  //  Entry Rule -- Buy Tuesday
XR1 = DayOfWeek() == 3;  //  Exit Rule -- Sell Wednesday
Buy = ER1;
Sell = XR1;
```

Listing 7.1 -- Enter Market on Close

The first parameter to SetTradeDelays is the number of bars to wait between signal and the Buy action—the entry to long positions. You want no delay. The second parameter is the delay between the signal and Sell. It is not necessary that it matches the delay for the buy, but it typically does.

The entry price is stored in the AmiBroker system variable named BuyPrice. In this example, it is the Close.

Remember that AmiBroker has a hierarchy for setting system parameters. Settings for BuyPrice, BuyDelay, etc, made in the Backtester Settings tabs determine the defaults. Those will be used unless you override them explicitly by having lines in your afl resetting them to your preference. I recommend always coding your preference into your afl.

Invariably, someone will point out that many of the signals depend on the closing price, and it is not possible to trade at the closing price if the close has already happened. *Quantitative Trading Systems* discusses this in more detail, but there are several ways to accomplish trading at the close of the action bar:

- Use a real-time data feed, note the price just before the close, perform the calculations, enter a market order or a limit-on-close order.
- Wait for the close, perform the calculations, trade in the after-hours market. Many ETFs and surrogates for them trade 15 minutes longer than regular hours.
- Pre-compute the price at which a signal will be generated. Depending on the data required for the calculation, this can be done just before the close, or as early as the previous evening if you are using daily bars. Place conditional orders to be executed if the closing price satisfies the conditions. See the section on Anticipating Action below.

It is easy to get confused about what is known at what time, and what

Next Day Open

Decide the time frame of reference for your system.

If the frame of reference is after the close of the current bar and before the open of the next bar, as it would be if you were updating data and running your system in the evening, the AmiBroker code is:

```
//   Enter Next Day Open
SetTradeDelays( 1, 1, 0, 0 );
BuyPrice = Open;
SellPrice = Open;
ER1 = DayOfWeek() == 2;   //   Entry Rule -- Buy Tuesday
XR1 = DayOfWeek() == 3;   //   Exit Rule -- Sell Wednesday
Buy = ER1;
Sell = XR1;
```

Listing 7.2 -- Enter Next Day Open Version 1

The code shown has a one bar delay between the signal and the Buy, with the Buy taking place at the open of the next bar. The entry price is the Open. The entry rule is to buy Tuesday. The rule is True on Tuesday. The entry happens at Wednesday's open.

The code above also has a one bar delay for the sell. That is not a requirement. The system could buy at the open, then sell intra-day or at the close of the same bar, as the next example does. In this case, the delay for the sell would be zero. In order for this to work, be certain to check the box "Allow same bar exit" on Backtester Settings > General tab.

```
//   Enter Next Day Open
SetTradeDelays( 1, 0, 0, 0 );
BuyPrice = Open;
SellPrice = Close;
ER1 = DayOfWeek() == 2;   //   Entry Rule -- Buy Tuesday
XR1 = DayOfWeek() == 3;   //   Exit Rule -- Sell Wednesday
Buy = ER1;
Sell = XR1;
```

Listing 7.3 -- Enter Next Day Open Version 2

If the frame of reference is at or shortly after the next bar has opened, the AmiBroker code is:

```
//   Today's Open, referencing yesterday
SetTradeDelays( 0, 0, 0, 0 );
BuyPrice = Open;
SellPrice = Open;
ER1 = DayOfWeek() == 2;   //   Entry Rule -- Buy Tuesday
XR1 = DayOfWeek() == 3;   //   Exit Rule -- Sell Wednesday
Buy = Ref( ER1, -1 );
Sell = Ref( XR1, -1 );
```

Listing 7.4 -- Enter Next Day Open Version 3

Indicators and signals are computed normally, but the reference to them is delayed in the Buy statement. Ref (ER1, -1) is True if ER1 was True upon completion of the previous bar. ER1 is True on Tuesday, so this code enters a long position at the Open on Wednesday and exits on Thursday.

LIMIT ORDER

A limit order to enter a long position is placed with a limit price below the current price.

If you are computing the price after the close of one bar for possible execution at or following the open of the next bar, you need a statement something like:

```
LimitPrice = 0.99 * Close;
```

to establish the price (1% below the previous Close) so you can enter the order with your broker. You will not know whether it was filled until there is trading in the next bar—perhaps not until the end of the next bar.

Again, you have a choice for your frame of reference. To my thinking, the program is clearer and less likely to develop an unintended future leak if you compute the indicators and signals normally, then compare their previous values to the current bar's data to test whether the order was filled:

```
//   Limit Order
SetTradeDelays( 0, 0, 0, 0 );
//   Compute LimitPrice today for use tomorrow
LimitPrice = 0.99 * Close;
//   Refer to yesterday's LimitPrice
BuyPrice = Ref( LimitPrice, -1 );
SellPrice = Close;
Buy = L < Ref( LimitPrice, -1 );
Sell = 1;
```

Listing 7.5 -- Enter Using Limit Order

Stop Order

Again, I recommend writing the code assuming the order will be filled in the current bar:

```
//   Stop Order
SetTradeDelays( 0, 0, 0, 0 );
Slippage = 0.02; //   Your best estimate
//   Compute StopPrice today for use tomorrow
StopPrice = 1.01 * H;
//   Refer to yesterday's StopPrice
BuyPrice = Ref( StopPrice, -1 ) + Slippage;
SellPrice = Close;
Buy = H > Ref( StopPrice, -1 );
Sell = 1;
```

Listing 7.6 -- Enter Using Stop Order

Future Leak

Be aware of the possibility of creating a *future leak* by using information not available in real trading at the time of the computation of an indicator or signal. Using code with known future leaks can be very valuable in research studies, but cannot be used to trade.

The following code will only enter a long position when it is assured of a profitable exit.

```
//   Future leak
//   Enter at the Open, knowing the Close
SetTradeDelays( 0, 0, 0, 0 );
BuyPrice = Open;
SellPrice = Close;
ER1 = DayOfWeek() == 2 AND C > O;
XR1 = 1;
Buy = ER1;
Sell = XR1;
```

Listing 7.7 -- Future Leak

Some future leaks are detected by the Check feature of the Editor. It does not detect this one. It has difficulty detecting future leaks in looping code, and it cannot detect future leaks in dlls.

Everyone has an unintended future leak or two creep into their code at some time. A symptom that your code has one is that your backtest results look too good to be true.

ANTICIPATING ACTION

It is often advantageous to know the price at which some condition will be true. For some simple indicators, such as the crossover of two simple moving averages, a little algebra will produce a formula allowing direct computation of the desired value. Although most indicators cannot be "reverse engineered" in this way, it is still possible to compute the price if some conditions are met.

- The solution must depend on a single value. You can solve for the price that gives an RSI of a certain level, because it uses only a single price series. But you cannot solve for a DV2 reaching a certain level, because it uses both high and low. (If it is near the end of the day, and you are reasonably certain that one of the high or low will not change, you can enter the assumed value for one and solve for the other.)
- The solution must be unique. You can solve for the low that gives z-score of -1.0. There will be only one price that satisfies that. You can solve for RSI of 20. You cannot solve for RSI of 0, because many values result in RSI being 0. (You can get close by solving for RSI of 2 and RSI of 1, then extrapolating.)
- AmiBroker must be able to evaluate the formula.

The technique is based on root finding methods. Details, including AmiBroker code, can be found in *Quantitative Trading Systems*.

This is a somewhat advanced technique, and you will need to be able to program AmiBroker to implement it. It is worth learning. You will be able to develop systems that enter intra-day on limit orders.

8

Controlling Risk

The potential growth of your trading account is determined by the characteristics of the trading system and your personal risk tolerance.

Many things we do in life are limited by the risk we are willing to accept—from physical activities such as whether to scuba dive or how steep a ski run to take, to financial activities such as how aggressive to be in trading.

Your personal risk tolerance determines the maximum position size that can be used. It is vitally important that you:

- Understand that backtest results, without further validation, cannot be used to estimate future performance. Even walk forward results always overestimate profit potential and underestimate risk.
- Use a realistic set of trades as your "best estimate" of future performance as you perform the risk analysis to determine position size.
- Make conservative evaluations of your own risk tolerance. Think through possible scenarios, including market conditions worse than any you can remember, and decide what level of drawdown you are willing to accept.
- Understand that position size is dynamic. As system performance changes, position size must also change.
- Understand that placing trades larger than the maximum position size determined through the Monte Carlo simulation of future performance will lead to deeper drawdowns than you are willing to accept.
- Understand that all trading systems have the potential to lose the

entire trading account, no matter how carefully the validation and analysis have been performed.

Techniques to assess system health, risk, profit potential, and position size are discussed in *Modeling Trading System Performance*. Simulation tools that operate as an Excel add-in are provided and completely documented. This chapter assumes that you are familiar with those techniques and tools.

In this chapter, three methods of reducing risk are discussed:
- Use filters to help identify and avoid high risk conditions.
- Understand the effect of position size and adjust it in accordance with risk and system performance.
- Take positions in options rather than in the futures, index, or ETF the system is based on.

Filter—ATR

ATR—Average True Range—is a measure of volatility of prices. The average true range for a single day/bar is the difference between two values:
- The true-high, which is the greater of the high of the current bar and the close of the previous bar.
- The true-low, which is the lesser of the low of the current bar and the close of the previous bar.

ATR is typically smoothed over a lookback window, LB. Using daily data, ATR(LB) is the average of single day ATR values for LB days.

The program in Listing 8.1 is well known to be a profitable mean reversion system. It will be used to test the effectiveness of the ATR filter. It takes a long position when RSI(2) falls below 25 and exits when RSI(2) rises above 75.

```
//   ATRFilter.afl
//
//   Test a range of settings of ATR length
//   and ATR level to see if a mean reversion
//   system benefits from using it as a filter
//   to allow or block trades.
//

SetOption( "Initialequity", 100000 );
MaxPos = 1;
SetOption( "MaxOpenPositions", MaxPos );
SetPositionSize( 10000, spsValue );

SetOption( "ExtraColumnsLocation", 1 );

ATRLB = Optimize( "ATRLB", 8, 2, 10, 1 );
ATRValueLowerLimit = Optimize( "ATRValueLowerLimit", 60, 0, 95, 5 );

ATRValue = ATR( ATRLB );

ATRValuePR = PercentRank( ATRValue, 100 );

ATRFilter = ATRValuePR >= ATRValueLowerLimit
            AND ATRValuePR <= ATRValueLowerLimit + 5;

RSI2 = RSI( 2 );

Buy = RSI2 < 25;
//Buy = ATRFilter AND RSI2 < 25;
Sell = RSI2 > 75;

//   Plots

Plot( C, "C", colorBlack, styleCandle );
//Plot( RSI2, "RSI2", colorRed, styleLine | styleOwnScale );
//Plot( ATRValue, "ATRValue", colorGreen, styleLine | styleOwnScale );
Plot( ATRValuePR, "ATRValuePR", colorBlue, styleLine | styleOwnScale );

////////////// end      //////////////
```

Listing 8.1 -- RSI(2) for Filter Testing

There are two Buy statements. One has no filter, the other—commented out in Listing 8.1—requires that conditions related to the value of ATR satisfy a filter in order for the trade to be taken.

Mean Reversion Trading Systems

BASE RESULTS—NO FILTER

If there is no filter, that system makes 254 trades over the 13 year test period, gaining $11,935. Figure 8.1 shows the equity and drawdown curve and Figure 8.2 shows the trading statistics.

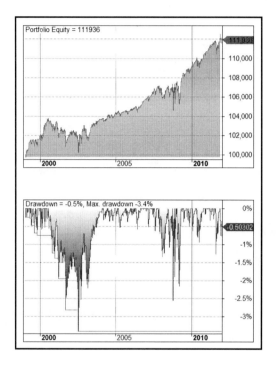

Figure 8.1 -- Equity and Drawdown -- No Filter

Statistics			
	All trades	Long trades	Short trades
Initial capital	100000.00	100000.00	100000.00
Ending capital	111935.80	111935.80	100000.00
Net Profit	11935.80	11935.80	0.00
Net Profit %	11.94 %	11.94 %	0.00 %
Exposure %	4.41 %	4.41 %	0.00 %
Net Risk Adjusted Return %	270.55 %	270.55 %	N/A
Annual Return %	0.87 %	0.87 %	0.00 %
Risk Adjusted Return %	19.75 %	19.75 %	N/A
All trades	254	254 (100.00 %)	0 (0.00 %)
Avg. Profit/Loss	46.99	46.99	N/A
Avg. Profit/Loss %	0.47 %	0.47 %	N/A
Avg. Bars Held	7.04	7.04	N/A
Winners	180 (70.87 %)	180 (70.87 %)	0 (0.00 %)
Total Profit	31597.52	31597.52	0.00
Avg. Profit	175.54	175.54	N/A
Avg. Profit %	1.76 %	1.76 %	N/A
Avg. Bars Held	4.97	4.97	N/A
Max. Consecutive	13	13	0
Largest win	1082.69	1082.69	0.00
# bars in largest win	6	6	0
Losers	74 (29.13 %)	74 (29.13 %)	0 (0.00 %)
Total Loss	-19661.72	-19661.72	0.00
Avg. Loss	-265.70	-265.70	N/A
Avg. Loss %	-2.66 %	-2.66 %	N/A
Avg. Bars Held	12.07	12.07	N/A
Max. Consecutive	3	3	0
Largest loss	-1450.86	-1450.86	0.00
# bars in largest loss	15	15	0
Max. trade drawdown	-2728.81	-2728.81	0.00
Max. trade % drawdown	-26.77 %	-26.77 %	0.00 %
Max. system drawdown	-3527.13	-3527.13	0.00
Max. system % drawdown	-3.40 %	-3.40 %	0.00 %
Recovery Factor	3.38	3.38	N/A
CAR/MaxDD	0.26	0.26	N/A
RAR/MaxDD	5.81	5.81	N/A
Profit Factor	1.61	1.61	N/A
Payoff Ratio	0.66	0.66	N/A
Standard Error	980.60	980.60	0.00
Risk-Reward Ratio	0.79	0.79	N/A
Ulcer Index	0.75	0.75	0.00
Ulcer Performance Index	-6.02	-6.02	N/A
Sharpe Ratio of trades	0.68	0.68	0.00
K-Ratio	0.0519	0.0519	N/A

Figure 8.2 -- Statistics -- No Filter

FILTER RESEARCH

To test the effect of ATR on system performance, the Buy statement that includes the filter is used. The percent rank function has been used to transform ATR into a linear-like distribution with values between 0 and 100. Note the two Optimize statements. The first sets the length of the ATR lookback window to values of 2 through 10; the second sets the lower limit for a window that creates 20 vigintiles of ATR values. A total of 180 tests are made. 20 with ATR of length 2, 20 with ATR of length 3, and so forth through 20 with ATR of length 10. For a given length of ATR window, there are 20 test results—one for each vigintile of the value of the ATR. The lowest vigintile includes all trades made when the ATR value was between 0 and 5. The next lowest collects trades made when the ATR was between 5 and 10. And so on.

The results were exported to Excel, where they were sorted and summarized using a pivot table. See Figure 8.3. Each column is an ATR lookback length; each row is an ATR vigintile; the number in the cell is the net profit for that combination. Row 60, column 8, contains 4443. When the ATR has a lookback length of 8, and the values are in the vigintile between 60 and 65, profit is $4,443. Setting ATR lookback to 8, then making 20 separate tests, one for each 5 point range of ATR values, gives the profits that are listed in column 8. They add up to $29,265. Setting ATR lookback length to 8, then making a single test where all trades were permitted gives a profit of $11,935. The difference between $29,265 and $11,935 is, as they say, rounding error. It is due to complications of trades at the boundaries, and the fact that signals are taken as levels rather than as impulses.

Sum of Net Profit	Column Labels									
Row Labels	2	3	4	5	6	7	8	9	10	Grand Total
0	1225	1426	1949	2368	2714	2488	2145	1725	1869	17908
5	1462	474	379	410	517	269	-1503	-2255	-2068	-2315
10	20	-474	1091	1199	722	-1347	-526	-1285	-561	-1162
15	1921	1677	-422	-1969	-1251	-2893	-1978	-845	1680	-4080
20	-1141	-3735	-1464	-2830	-3131	-1228	-810	-895	66	-15169
25	-431	143	-4616	-1615	-1410	-1350	-837	-1019	-1281	-12415
30	-864	-4757	2392	1486	2396	1780	1465	1185	-30	5053
35	-3264	2275	2503	1357	273	-551	-239	-26	941	3269
40	-3373	1833	817	933	728	-753	1077	392	909	2563
45	2075	2061	1518	1742	1935	3195	3008	1692	2252	19478
50	1726	1328	1707	1862	2905	3104	3084	2551	1899	20167
55	3955	2463	1749	3286	4698	3015	1765	3507	4571	29009
60	1239	2953	2308	3699	4555	3597	4443	2167	1459	26420
65	4125	5117	5130	3250	1789	2809	1714	2859	2298	29090
70	4884	4181	4743	4562	4549	2375	2123	1133	1594	30144
75	4997	5447	3284	2758	714	3024	2403	4573	2230	29429
80	5290	2623	2920	1834	1353	3426	2465	1295	2068	23274
85	3633	6618	5210	3646	5130	5091	3707	4237	3745	41017
90	3900	3577	4218	3981	2861	3943	2920	2626	3121	31147
95	2658	3290	3885	4244	3827	3047	2840	3236	3850	30878
Grand Total	34039	38520	39301	36202	35874	33042	29265	26852	30612	303707

Figure 8.3 -- Profit Related to ATR Lookback and ATR Value

Controlling Risk

Figure 8.4 shows the plot of profit by ATR lookback and ATR value. The height of each column represents the profit of trades made with that combination of ATR lookback and ATR value.

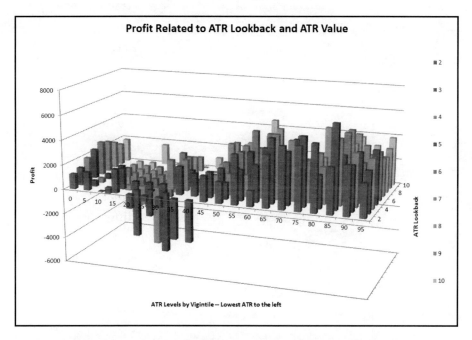

Figure 8.4 -- Plot of Profit Related to ATR Lookback and ATR Value

Assuming the relationship holds over all 13 years of the test:
- ATR lookback is relatively unimportant.
- ATR value is very important.

Trades are much less profitable when ATR value is in the range of 5 to 45. More research is needed to know whether very low levels (the left-most vigintile) of ATR are profitable or not. But it does seem to be reasonable to require ATR level to be 50 or higher to permit trades. The code segment to do that is shown in Listing 8.2. The variable named ATRFilter is true only when ATRValuePR is 50 or greater. This represents the entire right-hand side (half, in this case) of the distribution.

```
ATRFilter = ATRValuePR >= 50;
RSI2 = RSI( 2 );
//Buy = RSI2 < 25;
Buy = ATRFilter AND RSI2 < 25;
Sell = RSI2 > 75;
```

Listing 8.2 -- Set a Filter Based on ATR Value

The resulting system takes 151 trades and gains $12,387. Figure 8.5 shows the equity and drawdown curves. Note the smoother equity curve throughout, and the lower drawdown, particularly in the period 2000 to 2004, 2008, and 2009.

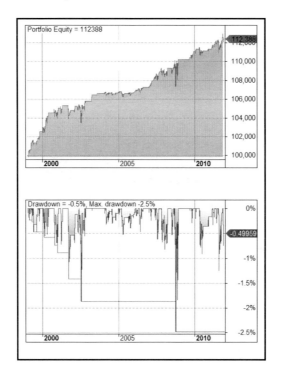

Figure 8.5 -- Equity and Drawdown using ATR Filter

Controlling Risk

FILTERED RESULTS

Figure 8.6 shows the statistics. The system using the filter took 40% fewer trades, had a higher percentage of winning trades, and lower system drawdown. It did not filter out the worst single trade. Average profit per trade for all trades rose from 0.47% to 0.82%.

Statistics	All trades	Long trades	Short trades
Initial capital	100000.00	100000.00	100000.00
Ending capital	112387.96	112387.96	100000.00
Net Profit	12387.96	12387.96	0.00
Net Profit %	12.39 %	12.39 %	0.00 %
Exposure %	2.48 %	2.48 %	0.00 %
Net Risk Adjusted Return %	500.41 %	500.41 %	N/A
Annual Return %	0.90 %	0.90 %	0.00 %
Risk Adjusted Return %	36.47 %	36.47 %	N/A
All trades	151	151 (100.00 %)	0 (0.00 %)
Avg. Profit/Loss	82.04	82.04	N/A
Avg. Profit/Loss %	0.82 %	0.82 %	N/A
Avg. Bars Held	6.79	6.79	N/A
Winners	110 (72.85 %)	110 (72.85 %)	0 (0.00 %)
Total Profit	21999.20	21999.20	0.00
Avg. Profit	199.99	199.99	N/A
Avg. Profit %	2.00 %	2.00 %	N/A
Avg. Bars Held	5.22	5.22	N/A
Max. Consecutive	13	13	0
Largest win	1082.69	1082.69	0.00
# bars in largest win	6	6	0
Losers	41 (27.15 %)	41 (27.15 %)	0 (0.00 %)
Total Loss	-9611.24	-9611.24	0.00
Avg. Loss	-234.42	-234.42	N/A
Avg. Loss %	-2.34 %	-2.34 %	N/A
Avg. Bars Held	11.02	11.02	N/A
Max. Consecutive	3	3	0
Largest loss	-1450.86	-1450.86	0.00
# bars in largest loss	15	15	0
Max. trade drawdown	-2728.81	-2728.81	0.00
Max. trade % drawdown	-26.77 %	-26.77 %	0.00 %
Max. system drawdown	-2728.81	-2728.81	0.00
Max. system % drawdown	-2.48 %	-2.48 %	0.00 %
Recovery Factor	4.54	4.54	N/A
CAR/MaxDD	0.36	0.36	N/A
RAR/MaxDD	14.72	14.72	N/A
Profit Factor	2.29	2.29	N/A
Payoff Ratio	0.85	0.85	N/A
Standard Error	673.32	673.32	0.00
Risk-Reward Ratio	1.11	1.11	N/A
Ulcer Index	0.28	0.28	0.00
Ulcer Performance Index	-16.21	-16.21	N/A
Sharpe Ratio of trades	1.61	1.61	0.00
K-Ratio	0.0730	0.0730	N/A

Figure 8.6 -- Statistics using ATR Filter

Is ATR an Effective Filter?

To determine whether ATR is an effective filter, we need to look at the performance of the two alternatives, normalized for risk. One alternative is the unfiltered system, the other requires Percent Rank of ATR(8) to be 50 or greater.

The set of trades produced by the backtest of each will be used as its best estimate set. Since these are backtest results, they have limited value in estimating future performance. But since the primary difference between them is whether there is a filter or not, we will use them as the basis for comparison. Each represents the same 13 year test period. The unfiltered version has 254 trades—about 19.5 per year. The filtered version has 151 trades—about 11.6 per year. Figure 8.7 shows the trades of each, sorted by percentage gained per trade.

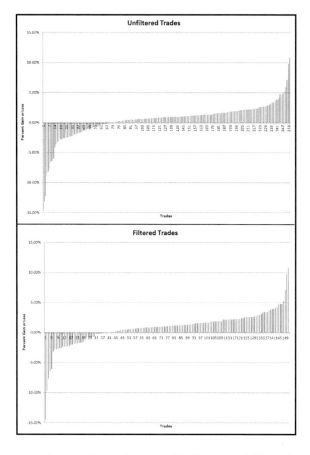

Figure 8.7 -- Trades from Unfiltered and Filtered

Controlling Risk

Notice that although the very largest losing trade was not avoided, most of the trades that lost more than 5% were avoided. It is losing trades that are the primary cause of risk of drawdown.

The forecast horizon is arbitrarily chosen to be two years. In two years, the unfiltered system will have an average of 39 trades; the filtered system an average of 23 trades.

Two separate simulations are run using the techniques described in *Modeling Trading System Performance*. One to estimate the risk of the unfiltered system; the second to estimate the risk of the filtered system.

A measure of risk is a statement about both the depth of drawdown that is acceptable and the certainty that depth will not be exceeded.

Each person has his or her own risk tolerance. The definition of risk tolerance used for this comparison, risk tolerance is: we want 90% certainty that the maximum drawdown will not exceed 20%. Maximum drawdown is measured from highest equity to date. The initial account balance is $100,000. Figure 8.8 shows a summary of the two systems.

	Unfiltered	Filtered
Number Trades	39	23
Average Profit / Trade	0.48%	0.87%
Maximum Fraction	0.85	1.25
CAR 10	-2.77%	-0.83%
CAR 50	7.74%	11.61%
CAR 90	17.38%	23.98%
Max DD 10	5.07%	2.99%
Max DD 50	10.43%	8.03%
Max DD 90	19.77%	19.90%

Figure 8.8 -- Comparison of Performance for Equal Risk

Maximum fraction is the maximum fraction of the account that could be used for each trade while holding maximum drawdown to 20% with 90% certainty. Unfiltered maximum fraction is 0.85—meaning 85% of the account balance. Using a higher position size increases the expected amount of drawdown and the likelihood that a larger drawdown will occur. Maximum fraction for the filtered system is 1.25. This version can be traded using 25% additional leverage—either through use of margin or ETFs with beta greater than 1.

The mean CAR—Compound Annual Return—at maximum fraction is 7.74% for the unfiltered version; 11.61% for the filtered version. Additionally (see Figure 8.9) the unfiltered version has a much higher probability that it will not be profitable over any random two year period.

The mean Max DD—the average maximum drawdown of all 1000 simulations—is 10.43% for the unfiltered version and 8.03% for the filtered version. Note the row for Max DD 90—that is the 90% certainty row. When both systems have the same value in that row, 20% in this case, risk is normalized allowing comparison of the other metrics.

Figure 8.9 shows the distribution of net profit per trade.

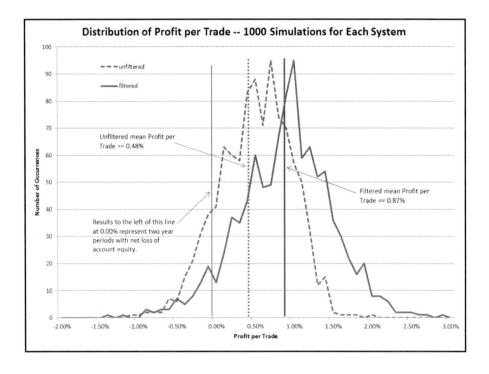

Figure 8.9 -- Distribution of Net Profit per Trade

The dotted line represents the unfiltered system, the solid line represents the filtered system. There are three vertical lines—one at 0.48%, the mean profit per trade of the unfiltered system; one at 0.87%, the mean of the filtered system; and one at 0.00%, the dividing line between being profitable or not after two years.

Conclusion

Based on the results shown in the table in Figure 8.8 and visual inspection of Figure 8.9, the performance of the filtered system is superior to that of the unfiltered system. But a caution is in order. Even though the equity curves in both Figure 8.1 and Figure 8.5 look smooth and safe, and there is significant improvement through use of the filter, there is still risk. There is a 17% probability that the unfiltered system will show a loss over any two year period, and maximum drawdown at the 99th percentile could reach 29%. The filtered system is better, but there is an 8% probability of no profit and a maximum drawdown at the 99th percentile of 33%. (The primary reason the maximum drawdown is higher for the filtered system is that it has so many fewer trades.)

Regardless of the logic used, all mean reversion systems based on daily data are trying to identify the same conditions—prices are over-extended and likely to return to a mean within a few days. So, results of testing the ATR filter on the RSI25 system may suggest a generalization.

Mean reversion systems work best in periods of high volatility.

By all means, do your own research using your own system.

Filter—Moving Average

The same technique can be used to test the position of the most recent closing price relative to a moving average as filter. Traditional wisdom suggests avoiding long trades when the close is below the, say, 200 day moving average. The program used to test ATR is modified slightly and used to test moving average. Listing 8.3 has the AmiBroker code.

```
//    MovingAverageFilter.afl
//
//    Test a range of settings of moving average length.
//    Require the closing price be above the moving
//    average to allow the trade.
//

//   System settings

SetOption( "Initialequity", 100000 );
MaxPos = 1;
SetOption( "MaxOpenPositions", MaxPos );
SetPositionSize( 10000, spsValue );

SetOption( "ExtraColumnsLocation", 1 );

SetTradeDelays( 0, 0, 0, 0 );
BuyPrice = SellPrice = Close;
```

```
//   User functions

function ComputeFib( n )
{
//   The function ComputeFib accepts an integer, n,
//   computes the nth Fibonacci number, and returns
//   that value.

//   n    Fib
//   1    1
//   2    2
//   3    3
//   4    5
//   5    8
//   6    13
//   7    21
//   8    34
//   9    55
//   10   89
//   11   144
//   12   233

    if ( n <= 1 )
        Fib = 1;
    else
    {
        f[0] = 1;
        f[1] = 1;

        for ( j = 2; j <= n; j++ )
        {
            f[j] = f[j-1] + f[j-2];
        }

        Fib = f[n];
    }

    return ( Fib );
}

//   Parameters

//   To cover a wide range of moving average lengths
//   without performing a lot of test runs,
//   the Fibonacci numbers are used as lookback lengths.

Fibn = Optimize( "Fibn", 6, 1, 12, 1 );
LB = ComputeFib( Fibn );
FilterMA = MA( C, LB );

//   Indicators

RSI2 = RSI( 2 );
FilterPass = C >= FilterMA;

//   Signals

Buy = RSI2 < 25 AND FilterPass;
```

Controlling Risk

```
Sell = RSI2 > 75;

// Plots

Plot( C, "C", colorBlack, styleCandle );
Plot( FilterMA, "FilterMA", colorGreen, styleLine );

//////////////////   end   //////////////////
```

Listing 8.3 -- Test Moving Average as Filter

When run as an optimization, 12 moving average lengths were computed and each was used as a filter. Figure 8.10 shows a table with the summary results for those 12 runs. The length of the moving average is in the third column.

Net Profit		Length	CAR	Profit Fact...	Avg % Profit/Lo...	K-Ratio	% of Winn...	Avg Bars Held
11,935.80	1	1	0.09	1.61	0.47	0.0519	70.87	7.04
3,482.06	2	2	0.03	3.16	0.81	0.0488	72.09	5.81
402.11	3	3	0.00	N/A	1.01	0.0382	100.00	4.00
0.00	4	5	0.00	N/A	N/A	N/A	N/A	N/A
1,507.89	5	8	0.01	2.51	0.89	0.0220	82.35	4.71
7,536.88	6	13	0.06	2.92	0.84	0.0879	77.78	6.24
7,833.23	7	21	0.06	2.20	0.63	0.0567	75.00	6.76
7,306.26	8	34	0.06	1.91	0.50	0.0468	73.79	6.80
7,574.20	9	55	0.06	2.04	0.51	0.0433	73.83	6.92
8,246.92	10	89	0.06	2.41	0.56	0.0508	74.83	6.77
7,086.51	11	144	0.05	2.13	0.49	0.0408	72.41	6.79
8,400.07	12	233	0.06	2.36	0.55	0.0884	72.37	6.63

Figure 8.10 -- Summary of a Range of Moving Average Lengths

The highest profit came with the moving average length of 1. Since the close is always equal to the 1-period moving average, there is no filter effect. Using K-Ratio as the metric, lengths 13 and 233 were highest. Twelve separate backtest runs were made—one with each of the lengths. Figure 8.11 shows all 12.

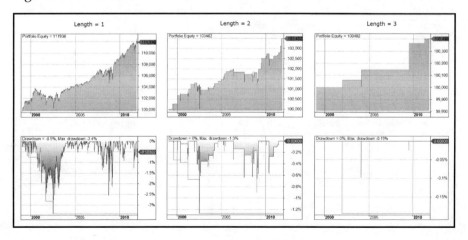

Mean Reversion Trading Systems

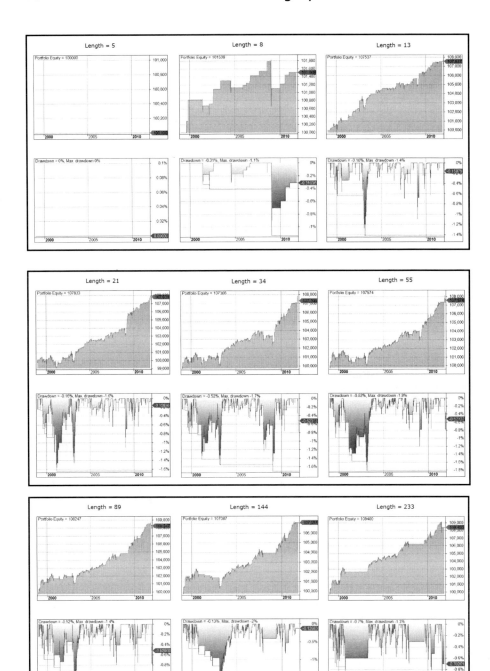

Figure 8.11 -- Equity Curves for 12 Moving Average Filters

Controlling Risk

The vertical scale was determined automatically by the reporting program. Although all 12 appear to rise smoothly to the same level, the net profit is highest for length 1 ($11,935), low for lengths 2, 3, 5, and 8, then roughly the same ($7,000 to $8,400) for lengths 13 through 233. There were no trades for length 5. The primary cycle length is about five bars, so there were no days when the price was above the 5 day average and 2 period RSI was low.

The effectiveness of using a moving average as a filter may be different for other trading systems. But the longer length moving averages keep the system flat for very long periods and forgo many profitable trades. Use your judgement with your system, but I think there are better filters than moving averages.

Position Size

As illustrated in the previous section, maintaining the proper position size is critical to managing risk and maximizing account growth.

Risk and associated drawdown is a function of losing trades—the number of losing trades and the size of the losses. Position size is a function of risk. Risk assessment and position size determination must be dynamic—it is not a component of the trading system, nor can it be determined one time and forgotten. As conditions change, system performance changes, and position size must be adjusted.

Alternative trading systems, such as a system without a filter and that same system with a filter, cannot be properly evaluated until the distribution of risk has been estimated. Then, based on normalized risk, the systems can be compared.

Please refer to *Modeling Trading System Performance* for a more thorough discussion of the techniques for estimating future trading system performance, and for instructions for using the platform-independent simulation tools included with the book at no additional cost.

Options

To study use of options as surrogates for the underlying, we use the naive system from Chapter 3. It trades SPY long / flat and produces 207 trades in the test period. The mean gain per trade is 0.47%. Trades are typically held 1 day; the average is 2.4 days; about 10 percent are held longer than 4 days; the longest holding is 12 days.

There are many option pricing formulas. Black-Scholes is probably the

best known and most easily programmed. Option prices depend on five variables:
- Price of the underlying.
- Strike price of the option.
- Volatility of the underlying.
- Time to expiration.
- Risk free return.

A spreadsheet was used to evaluate the options strategy, using data as follows:
- Price of the underlying. Date of the entry to and exit from each trade. Price of SPY at trade execution. Closing price is used. Trades are made market on close. SPY ranged between about 70 and 150 during the test period.
- Strike price of the option. SPY options have strikes every dollar. The first in-the-money strike is computed as the integer portion of the SPY price. If SPY is 110.35, 110 is taken to be the first strike ITM.
- Time to expiration. Monthly options expire the third Friday of every month. At entry to the trade, the option closest to expiration that had at least 10 days before expiration was chosen. Time to expiration ranged from 10 to 44 days, and averaged 25 days. If the trade would be held past expiration, it would be either closed early or rolled to the next month. In this test, there were no trades where this was necessary.
- Volatility. There are two measures of volatility. The first is the historical volatility of the underlying. It is measured by standard deviation of price of the underlying. The second is the estimate of future volatility—implied volatility. VIX is an index based on the implied 30-day volatility of the S&P 500 index, SPX. It is computed by the CBOE using real time bid and ask quotes of options on SPX, and is reported by most real-time data vendors.
  All SPY trades are market on close. The corresponding daily VIX closing prices are used to estimate volatility for calculation of option prices for our study.
- Risk free rate. The annualized risk free alternative—the 30 or 90 day government rate.

The Black-Scholes formula was used to calculate the price of the option at entry, and of that same option at exit. Several prices were spot-checked to assure the calculations were reasonably accurate and consistent. The result is a new list of 207 trades where the tradable is the front month first in-the-money call option.

Figures 8.12 and 8.13 show the distribution of trades using SPY and the distribution of trades using options, respectively. Each set of trades is sorted separately. The uniform trades for SPY in Figure 8.12 result from exits made at the profit target.

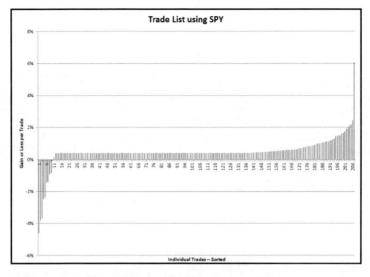

Figure 8.12 -- Trade List using SPY

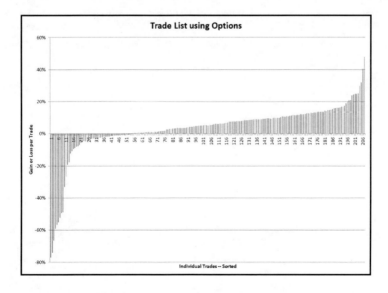

Figure 8.13 -- Trade List using Options

Figure 8.14 shows the scatter plot of the 207 trades. Gain or loss in SPY is on the horizontal axis, gain or loss in the corresponding option is on the vertical axis. A regression line has been fit to the data and shows a slope of about 15. Interpretation is that the average beta of the option relative to SPY is about 15.

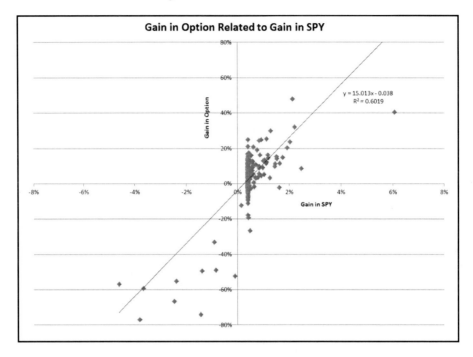

Figure 8.14 -- Gain in Option Related to Gain in SPY

A quick and dirty comparison of performance can be obtained by looking at the cumulative gain or loss of the trades. Figure 8.15 shows that with the solid line representing trades using SPY and the dotted lined for options.

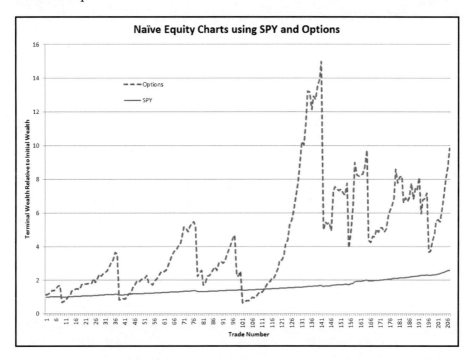

Figure 8.15 -- Naive Equity Curves using SPY and Options

A much better comparison of performance can be obtained by first analyzing each system separately to determine its risk and maximum position size, then comparing them based on normalized risk. The risk tolerance statement is the same used in the earlier example—for an account with a starting balance of $100,000 and a two year forecast horizon, we want 90% certainty that the maximum drawdown will not exceed 20%.

Trading SPY, the position size is the fraction of the account that can be used for each trade. After performing the analysis as described in *Modeling Trading System Performance*, the maximum fraction that can be used without exceeding the risk tolerance is 4.5. This system is a candidate to be traded using leveraged ETFs.

Trading options, there are two alternatives. The first treats the option the same as the underlying, and buys as many options as the fraction of the account can afford. The second treats options more like futures—it assumes a unit of trading, say $1,000, is a "contract" and buys the number of contracts in proportion to the account balance. For this example, the first method was used. The maximum fraction that can be used without exceeding the risk tolerance is 0.16. That is, only 16% of the account can be safely used for each trade.

Figure 8.16 shows a table comparing the results.

	SPY	Options
Number Trades	32	32
Average Profit / Trade	0.47%	3.19%
Standard Deviation / Trade	0.85%	16.53%
Maximum Fraction	4.5	0.16
CAR 10	20.05%	-2.88%
CAR 50	39.64%	8.67%
CAR 90	57.60%	18.12%
Max DD 10	0.00%	1.91%
Max DD 50	10.71%	10.67%
Max DD 90	20.75%	20.13%

Figure 8.16 -- Comparison of Performance for Equal Risk

The results using SPY are quite good. The standard deviation is less than two times the mean. (For most trading systems, the standard deviation is three or more times the mean.) Mean compound annual return is 39%, and there is less than 1% probability that there will be no net profit in any given two year period.

The results using options are quite poor. The standard deviation is over fives times the mean. Mean CAR is 8.7%.

Controlling Risk 169

Figure 8.17 shows the cumulative distribution of maximum drawdown for the two systems. Note the curves have the same value of 20% at the 90th percentile—that is where they are normalized.

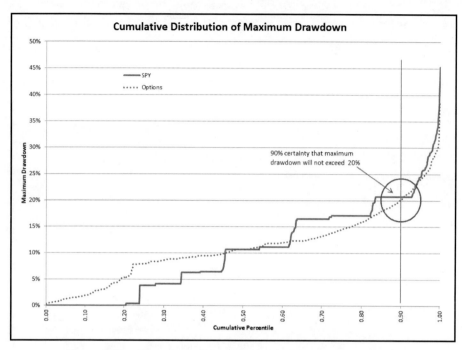

Figure 8.17 -- Cumulative Distribution of Maximum Drawdown

Two final figures show "straw broom" charts of ten equally likely two year equity curves. The vertical scales of the two charts are the same to help illustrate the difference between them.

Figure 8.18 shows the equity curves trading SPY using a fraction of 4.5. The heavy dotted line is the average of the ten—its final equity is about $191,000, consistent with CAR of 39%.

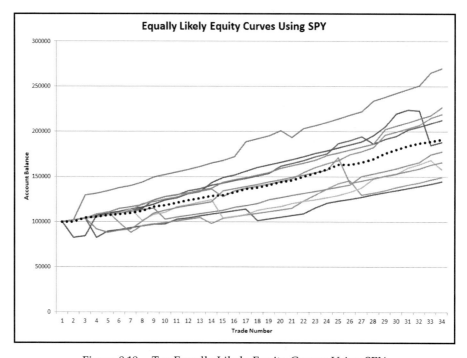

Figure 8.18 -- Ten Equally Likely Equity Curves Using SPY

Figure 8.19 shows the equity curves trading options using a fraction of 0.16. The heavy dotted lines is the average, with a final equity of about $127,000, a little higher than, but consistent with, CAR of 8.7%.

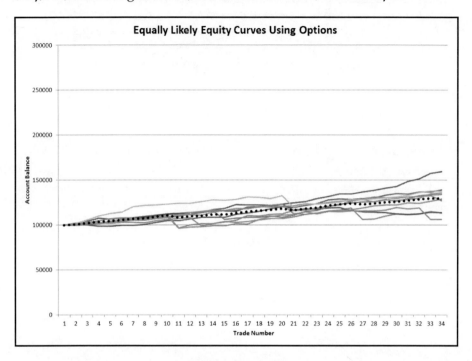

Figure 8.19 -- Ten Equally Likely Equity Curves Using Options

COMMENTARY

Many tests were run to determine the critical factors in using options as surrogates.
- Losing Trades. The system based on trading SPY has a profit target of 0.4% which was hit very often. This resulted in a high percentage of winning trades. Although there were not many losing trades, those that did lose lost a high percentage, particularly when using options. As we have found repeatedly—the number and size of losing trades is the most important limitation to a system.
- Time to Expiration. The time value component of the option price decays rapidly near expiration. There was little decay for those trades held one or two days, but significant decay for those held ten or more days. Systems that limit holding period to a few days are less susceptible to option premium decay.
- Volatility. Surprisingly, volatility was not a major factor. Tests

were run with higher volatility, lower volatility, volatility increasing during the trade, and decreasing during the trade—none had a significant effect.
- Risk Free Rate. Had essentially no effect.
- Dividends. Option prices are adjusted when the underlying pays a dividend. This is not an issue for SPY, but it might be when trading equities.

Conclusion

- Use of first strike in-the-money options with 10 to 40 days to expiration increases beta by a factor of about 15.
- Limit the holding period to one or two days to limit premium loss due to time decay.
- When using debit option positions, your maximum loss is known at trade entry.
- The number and size of losing trades continues to be the most significant factor in account growth.
- Inconsistency in the profit and loss of individual trades limits position size and, consequently, limits growth of the account.
- These results are backtest only, and for illustration. Perform your own validation and analysis before considering trading based on this system.

9

Systems

This chapter adds a few more examples of swing trading using mean reversion.

As a reminder—these system examples are intended to be educational and are not investment advice or recommendations for trading. Every individual has his or her preferences for data source, entry technique, exit technique, objective function, holding period, requirements for evidence of robustness, risk tolerance when determining position size, and many more variables. Before trading any of these systems, or any variation of them, you each must perform your own analysis. You must understand how they work and how they perform in all market conditions. You must have confidence in any system before you trade it.

Each system is illustrated with logic to trade long / flat. The system settings control initial equity and position size. All accounts begin at $100,000; all trades are a fixed size of $10,000. The test period is 1/1/1999 through 1/1/2012. The issue being traded is SPY.

Most of the systems have many potential variations that you can change as you wish:
- Using other indicators to identify being oversold.
- Adding additional positions if conditions become more oversold.
- Taking short trades.
- Applying different filtering techniques.
- Trading other issues.

The equity and drawdown curves shown with each system cover the same tradable issue, time period, and position sizing. They can be used for a quick comparison of the systems.

3 Day High Low System

The 3 day high low system comes from the book, "High Probability ETF Trading" by Larry Connors and Cesar Alvarez. There are two versions—basic and aggressive. The basic version takes a long position after a sequence of three lower days; or a short position after a sequence of three higher days. The aggressive version adds a second position. For long trades, the second position is added if the price closes below the first position's entry price; similarly for short trades. Listing 9.1 gives the AmiBroker code for the basic version for long trades.

```
//   3DayHighLow_Long.afl
//
//   Based on "High Probability ETF Trading"
//   by Larry Connors and Cesar Alvarez
//
//   3 Day High Low Method
//   Page 9
//
//   Basic rules for Long trades
//

//   System settings

OptimizerSetEngine( "cmae" );

SetOption( "Initialequity", 100000 );
MaxPos = 1;
SetOption( "MaxOpenPositions", MaxPos );
SetPositionSize( 10000, spsValue );

SetOption( "ExtraColumnsLocation", 1 );

SetTradeDelays( 0, 0, 0, 0 );
BuyPrice = SellPrice = Close;

//   User functions

//   Parameters

LongTermMALength = 200;
ShortTermMALength = 5;
ExitMALength = 5;

//   Indicators

LongTermMA = MA( C, LongTermMALength );
ShortTermMA = MA( C, ShortTermMALength );

//   Rules

//   Filter -- Close must be above its long term moving average
FilterRule1 = C > LongTermMA;

//   Close is below it short term moving average
```

```
EntryRule1 = C < ShortTermMA;
//   Three days of lower lows and lower highs
EntryRule2 = Ref( H, -2 ) < Ref( H, -3 )
             AND Ref( H, -1 ) < Ref( H, -2 )
             AND H < Ref( H, -1 )
             AND Ref( L, -2 ) < Ref( L, -3 )
             AND Ref( L, -1 ) < Ref( L, -2 )
             AND L < Ref( L, -1 );

//   Close is above its stort term moving average
ExitRule1 = C > ShortTermMA;

//   Signals

Buy = FilterRule1 AND EntryRule1 AND EntryRule2;
Sell = ExitRule1;

//   Plots

Plot( C, "Close", colorBlack, styleCandle );
Plot( LongTermMA, "LongTermMA", colorRed, styleLine );
Plot( ShortTermMA, "ShortTermMA", colorGreen, styleLine );

Shapes = IIf( Buy, shapeUpArrow,
         IIf( Sell, shapeDownArrow, shapeNone ) );
ShapeColors = IIf( Buy, colorGreen,
         IIf( Sell, colorRed, colorWhite ) );
PlotShapes( Shapes, ShapeColors );

////////////////////////////////

//   Exploration columns

/////////////////////// end ///////////////////////
```

Listing 9.1 -- 3 Day High Low System—Long Trades

Figure 9.1 shows the plot with the price as candlesticks, the long term moving average, short term moving average, and arrows when the buy and sell conditions are true. Note there are no buy arrows when the price is below the long term moving average.

Figure 9.1 -- Plot for Long Basic Trades

Figure 9.2 shows the equity and drawdown curves.

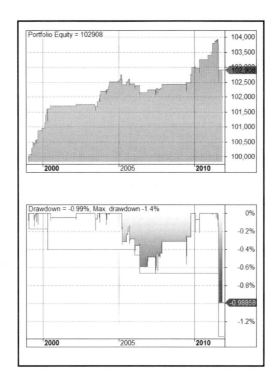

Figure 9.2 -- Equity and Drawdown for Long Basic Trades

Figure 9.3 shows the performance statistics.

Statistics		
	All trades	Long trades	Short trades
Initial capital	100000.00	100000.00	100000.00
Ending capital	102908.27	102908.27	100000.00
Net Profit	2908.27	2908.27	0.00
Net Profit %	2.91 %	2.91 %	0.00 %
Exposure %	0.46 %	0.46 %	0.00 %
Net Risk Adjusted Return %	638.49 %	638.49 %	N/A
Annual Return %	0.22 %	0.22 %	0.00 %
Risk Adjusted Return %	48.49 %	48.49 %	N/A
All trades	50	50 (100.00 %)	0 (0.00 %)
Avg. Profit/Loss	58.17	58.17	N/A
Avg. Profit/Loss %	0.58 %	0.58 %	N/A
Avg. Bars Held	4.06	4.06	N/A
Winners	41 (82.00 %)	41 (82.00 %)	0 (0.00 %)
Total Profit	4814.76	4814.76	0.00
Avg. Profit	117.43	117.43	N/A
Avg. Profit %	1.17 %	1.17 %	N/A
Avg. Bars Held	3.24	3.24	N/A
Max. Consecutive	10	10	0
Largest win	298.15	298.15	0.00
# bars in largest win	4	4	0
Losers	9 (18.00 %)	9 (18.00 %)	0 (0.00 %)
Total Loss	-1906.49	-1906.49	0.00
Avg. Loss	-211.83	-211.83	N/A
Avg. Loss %	-2.12 %	-2.12 %	N/A
Avg. Bars Held	7.78	7.78	N/A
Max. Consecutive	3	3	0
Largest loss	-1016.08	-1016.08	0.00
# bars in largest loss	12	12	0
Max. trade drawdown	-1404.29	-1404.29	0.00
Max. trade % drawdown	-14.04 %	-14.04 %	0.00 %
Max. system drawdown	-1415.70	-1415.70	0.00
Max. system % drawdown	-1.36 %	-1.36 %	0.00 %
Recovery Factor	2.05	2.05	N/A
CAR/MaxDD	0.16	0.16	N/A
RAR/MaxDD	35.60	35.60	N/A
Profit Factor	2.53	2.53	N/A
Payoff Ratio	0.55	0.55	N/A
Standard Error	339.04	339.04	0.00
Risk-Reward Ratio	0.51	0.51	N/A
Ulcer Index	0.29	0.29	0.00
Ulcer Performance Index	-17.82	-17.82	N/A
Sharpe Ratio of trades	2.03	2.03	0.00
K-Ratio	0.0336	0.0336	N/A

Figure 9.3 -- Statistics for Long Basic Trades

COMMENTARY

As this system is defined, it uses levels or states to indicate whether the system should be long or be flat. States change the value of a variable when the condition related to the state changes, and hold that new value until the state changes.

The alternative to states is impulses. Impulses are single-bar spikes that occur when some condition is met. Typically, the base value of the variable is zero, and the spike is 1—these correspond nicely with the numeric values associated with False and True, respectively.

In the chart in Figure 9.1, Buy has the value True, and there is an upward pointing green arrow below the price bars, for all bars where the three conditions are simultaneously true. A new long position is entered on the first occurrence of a series of green upward pointing arrows and held until the first occurrence of a red downward pointing arrow.

Variation—Trade Short / Flat

The 3 day system has a short / flat counterpart, with rules that are mirror images of the long / flat system. Listing 9.2 shows the AmiBroker code for the version that trades short / flat.

```
//    3DayHighLow_Short.afl
//
//    Based on "High Probability ETF Trading"
//    by Larry Connors and Cesar Alvarez
//
//    3 Day High Low Method
//    Page 9
//
//    Basic rules for Short trades
//

//  System settings

OptimizerSetEngine( "cmae" );

SetOption( "Initialequity", 100000 );
MaxPos = 1;
SetOption( "MaxOpenPositions", MaxPos );
SetPositionSize( 10000, spsValue );

SetOption( "ExtraColumnsLocation", 1 );

SetTradeDelays( 0, 0, 0, 0 );
ShortPrice = CoverPrice = Close;

//  User functions

//  Parameters

LongTermMALength = 200;
ShortTermMALength = 5;
ExitMALength = 5;

//  Indicators

LongTermMA = MA( C, LongTermMALength );
ShortTermMA = MA( C, ShortTermMALength );

//  Rules
```

Systems

```
//   Filter -- Close must be below its long term moving average
FilterRule1 = C < LongTermMA;

//   Close is above it short term moving average
EntryRule1 = C > ShortTermMA;
//   Three days of higher lows and higher highs
EntryRule2 = Ref( H, -2 ) > Ref( H, -3 )
             AND Ref( H, -1 ) > Ref( H, -2 )
             AND H > Ref( H, -1 )
             AND Ref( L, -2 ) > Ref( L, -3 )
             AND Ref( L, -1 ) > Ref( L, -2 )
             AND L > Ref( L, -1 );

//   Close is below its stort term moving average
ExitRule1 = C < ShortTermMA;

//   Signals

Short = FilterRule1 AND EntryRule1 AND EntryRule2;
Cover = ExitRule1;

//   Plots

Plot( C, "Close", colorBlack, styleCandle );
Plot( LongTermMA, "LongTermMA", colorRed, styleLine );
Plot( ShortTermMA, "ShortTermMA", colorGreen, styleLine );

Shapes = IIf( Cover, shapeUpArrow,
          IIf( Short, shapeDownArrow, shapeNone ) );
ShapeColors = IIf( Cover, colorGreen,
          IIf( Short, colorRed, colorWhite ) );
PlotShapes( Shapes, ShapeColors );


////////////////////////////////

//   Exploration columns

/////////////////////// end /////////////////////// s
```

Listing 9.2 -- 3 Day High Low System—Short Trades

Figure 9.4 shows the plot of price, the moving averages, and the short and cover arrows. Compare with Figure 9.1 to see how the long term moving average acts as a filter, allowing only long trades when the close is above it, and only short trades when the close is below it.

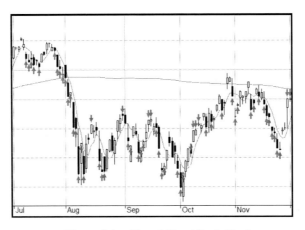

Figure 9.4 -- Plot of Short Basic Trades

Figure 9.5 shows the equity and drawdown curves.

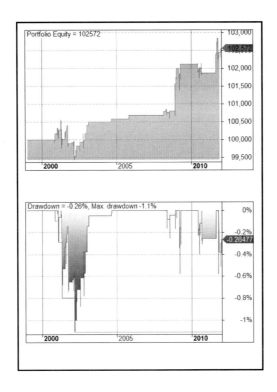

Figure 9.5 -- Equity for Short Basic Trades

Figure 9.6 shows the statistics.

Statistics	All trades	Long trades	Short trades
Initial capital	100000.00	100000.00	100000.00
Ending capital	102571.95	100000.00	102571.95
Net Profit	2571.95	0.00	2571.95
Net Profit %	2.57 %	0.00 %	2.57 %
Exposure %	0.49 %	0.00 %	0.49 %
Net Risk Adjusted Return %	527.97 %	N/A	527.97 %
Annual Return %	0.20 %	0.00 %	0.20 %
Risk Adjusted Return %	40.16 %	N/A	40.16 %
All trades	43	0 (0.00 %)	43 (100.00 %)
Avg. Profit/Loss	59.81	N/A	59.81
Avg. Profit/Loss %	0.60 %	N/A	0.60 %
Avg. Bars Held	4.77	N/A	4.77
Winners	33 (76.74 %)	0 (0.00 %)	33 (76.74 %)
Total Profit	4959.12	0.00	4959.12
Avg. Profit	150.28	N/A	150.28
Avg. Profit %	1.50 %	N/A	1.50 %
Avg. Bars Held	3.82	N/A	3.82
Max. Consecutive	13	0	13
Largest win	771.05	0.00	771.05
# bars in largest win	3	0	3
Losers	10 (23.26 %)	0 (0.00 %)	10 (23.26 %)
Total Loss	-2387.17	0.00	-2387.17
Avg. Loss	-238.72	N/A	-238.72
Avg. Loss %	-2.39 %	N/A	-2.39 %
Avg. Bars Held	7.90	N/A	7.90
Max. Consecutive	2	0	2
Largest loss	-506.95	0.00	-506.95
# bars in largest loss	9	0	9
Max. trade drawdown	-764.15	0.00	-764.15
Max. trade % drawdown	-7.64 %	0.00 %	-7.64 %
Max. system drawdown	-1113.92	0.00	-1113.92
Max. system % drawdown	-1.11 %	0.00 %	-1.11 %
Recovery Factor	2.31	N/A	2.31
CAR/MaxDD	0.18	N/A	0.18
RAR/MaxDD	36.25	N/A	36.25
Profit Factor	2.08	N/A	2.08
Payoff Ratio	0.63	N/A	0.63
Standard Error	345.85	0.00	345.85
Risk-Reward Ratio	0.54	N/A	0.54
Ulcer Index	0.25	0.00	0.25
Ulcer Performance Index	-21.01	N/A	-21.01
Sharpe Ratio of trades	1.81	0.00	1.81
K-Ratio	0.0354	N/A	0.0354

Figure 9.6 -- Statistics for Short Basic Trades

VARIATION—AGGRESSIVE MODE

This system has an aggressive mode. When the system is holding a single long position and the price closes lower than the entry price for the first position, buy a second position.

The program to keep track of the number of positions, the entry price for each, and the rules for adding and exiting positions, looks very much like a trading system development simulator. Each data bar must be processed in time order. Enough information must be stored to generate the necessary entry and exit signals. Each level of entry must be uniquely identified so that trades associated with that level can be evaluated on its own. The code shown in Listing 9.3 is a simplified simulator. It handles two levels of long positions according to the rules.

As the trading system simulator processes each bar of data, it does so in the order events would occur—at the open of the bar, intra-bar, at the close of the bar. Exits are made before entries, so funds and positions are immediately made available. If two events could both have taken place in the same processing period, assume the worst outcome. For example, if the low is low enough to trigger a maximum loss exit and the high is high enough to trigger a profit target in an intra-day period, record the exit that is least profitable. At the end of the bar, everything needed for processing the next bar must be up to date.

A rough outline is:
1. Initialize the simulator.
2. Using a loop, process all data bars:
   A. At the open:
      1) Process exits.
      2) Process entries.
      3) Update data.
   B. Intra-trade:
      1) Process exits, worst first.
      2) Process entries.
      3) Update data.
   C. At the close:
      1) Process exits.
      2) Process entries.
      3) Update data.
3. Decide which signals should be passed to AmiBroker.

```
//    SimulateMultiplePositions.afl
//
//    This program implements the 3 Day High Low
//    system for both basic and aggressive trades.
//    It allows either level of trades to be isolated
//    so they can be evaluated independently.
//
//    This is a simplified trading system simulator.
//

/*
The outline of simulator activities is:

1.    Initialize the simulator.
```

Systems

```
2.  Using a loop, process all data bars:
    A.  At the open:
        1)  Process exits.
        2)  Process entries.
        3)  Update data.
    B.  Intra-trade:
        1)  Process exits, worst first.
        2)  Process entries.
        3)  Update data.
    C.  At the close:
        1)  Process exits.
        2)  Process entries.
        3)  Update data.
3.  Decide which signals should be passed to AmiBroker.
*/

/*
Data is stored in arrays, one element for each data bar.
Nheld -- Number of long positions held
EntryP1 -- Entry Price for Position 1
EntryP2 -- Entry Price for Position 2
Buy1 -- Signal to enter the first position
Buy2 -- Signal to enter the Second position
SellAll -- Signal to exit all positions

for this Version, all actions take place at the Close.
Execution price is the closing price.

As much as can be done outside the loop
is done outside the loop.
*/


//  System settings

OptimizerSetEngine( "cmae" );

//  Allow multiple positions
SetBacktestMode( backtestRegularRawMulti );
MaxPos = 2;

SetOption( "Initialequity", 100000 );
SetOption( "MaxOpenPositions", MaxPos );
SetPositionSize( 10000, spsValue );

SetOption( "ExtraColumnsLocation", 1 );

SetTradeDelays( 0, 0, 0, 0 );
BuyPrice = SellPrice = Close;

//  User functions

//  Parameters

LongTermMALength = 200;
ShortTermMALength = 5;
ExitMALength = 5;
```

```
//    Indicators

LongTermMA = MA( C, LongTermMALength );
ShortTermMA = MA( C, ShortTermMALength );

//    Rules

//    Filter -- Close must be above its long term moving average
FilterRule1 = C > LongTermMA;

//    Close is below it short term moving average
EntryRule1 = C < ShortTermMA;
//    Three days of lower lows and lower highs
EntryRule2 = Ref( H, -2 ) < Ref( H, -3 )
             AND Ref( H, -1 ) < Ref( H, -2 )
             AND H < Ref( H, -1 )
             AND Ref( L, -2 ) < Ref( L, -3 )
             AND Ref( L, -1 ) < Ref( L, -2 )
             AND L < Ref( L, -1 );

//    Close is above its short term moving average
ExitRule1 = C > ShortTermMA;

////////////////////////////////
//    Trading system simulator loop
//    replaces these statements:
//
//    Signals
//
//Buy = FilterRule1 AND EntryRule1 AND EntryRule2;
//Sell = ExitRule1;
//
//

//    Initialize

Nheld = 0 * Close;
EntryP1 = 0 * Close;
EntryP2 = 0 * Close;
Buy1 = 0 * Close;
Buy2 = 0 * Close;
SellAll = 0 * Close;


//    Loop to process all data bars

for ( i = 1; i < BarCount; i++ )
{
//    At the open.
//    No actions at the open.

//    Intra-bar.
//    No actions intra-bar.

//    At the close.
//    Process Exits.
    if ( NHeld[i-1] > 0 && ExitRule1[i] )
```

```
            {
                SellAll[i] = 1;
                NHeld[i] = 0;
            }
            else
//          Process Entries
            switch ( NHeld[i-1] )
            {

                case 0:
                {
//              Holding none -- check for buy
                    if ( FilterRule1[i] AND EntryRule1[i] AND EntryRule2[i] )
                    {
//                  Enter a long position
                        Buy1[i] = 1;
                        EntryP1[i] = Close[i];
                        NHeld[i] = 1;
                    }
                    else
                    {
//                  No action
                        NHeld[i] = 0;
                    }
                    break;
                }

                case 1:
                {
//              Holding one -- check for aggressive entry
                    if ( Close[i] < EntryP1[i-1] )
                    {
//                  Take a second position
                        Buy2[i] = 1;
                        EntryP2[i] = Close[i];
                        NHeld[i] = 2;
                        EntryP1[i] = EntryP1[i-1];
                    }
                    else
                    {
//                  No action
                        NHeld[i] = 1;
                        EntryP1[i] = EntryP1[i-1];
                    }
                    break;
                }

                case 2:
                {
//              Holding two -- the maximum
//              No action
                    NHeld[i] = 2;
                    EntryP1[i] = EntryP1[i-1];
                    EntryP2[i] = EntryP2[i-1];
                    break;
                }
            }   //  end switch

    }   //  end of for i loop
```

```
//   Process Entries.
//   Pass signals to AmiBroker for reporting
TestingEntry = 2;     //   Trades at Level 1 or 2; 3 means both

switch ( TestingEntry )
{

case 1:
{
//   First buy
    Buy = Buy1;
    BuyPrice = EntryP1;
    Sell = SellAll;
    SellPrice = Close;
    break;
}

case 2:
{
//   Second buy -- aggressive
    Buy = Buy2;
    BuyPrice = EntryP2;
    Sell = SellAll;
    SellPrice = Close;
    break;
}

case 3:
{
//   Take all trades
    Buy = Buy1 + Buy2;
    BuyPrice = IIf(Buy1,EntryP1,IIf(Buy2,EntryP2,0));
    Sell = SellAll;
    SellPrice = Close;
    break;
}
}    //   end switch

//   End of simulator loop
////////////////////////////

//   Plots

Plot( C, "Close", colorBlack, styleCandle );

Plot( LongTermMA, "LongTermMA", colorRed, styleLine );

Plot( ShortTermMA, "ShortTermMA", colorGreen, styleLine );

Shapes = IIf( Buy, shapeUpArrow,
         IIf( Sell, shapeDownArrow, shapeNone ) );

ShapeColors = IIf( Buy, colorGreen,
         IIf( Sell, colorRed, colorWhite ) );

PlotShapes( Shapes, ShapeColors );
```

```
////////////////////////////
//    Exploration columns
/////////////////////  end  /////////////////////
```

Listing 9.3 -- 3 Day High Low System—Aggressive Trades

Figure 9.7 shows the arrows for the aggressive trades. All extraneous arrows have been removed.

Figure 9.7 -- Plot for Long Aggressive Trades

Figure 9.8 shows the equity and drawdown curves for long aggressive trades.

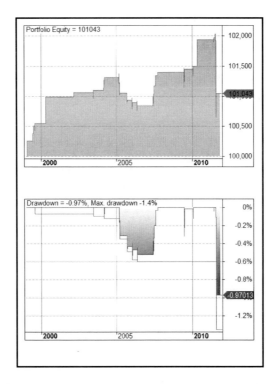

Figure 9.8 -- Equity for Long Aggressive Trades

Systems

Figure 9.9 shows the statistical summary for long aggressive trades.

Statistics	All trades	Long trades	Short trades
Initial capital	100000.00	100000.00	100000.00
Ending capital	101042.90	101042.90	100000.00
Net Profit	1042.90	1042.90	0.00
Net Profit %	1.04 %	1.04 %	0.00 %
Exposure %	0.23 %	0.23 %	0.00 %
Net Risk Adjusted Return %	462.78 %	462.78 %	N/A
Annual Return %	0.08 %	0.08 %	0.00 %
Risk Adjusted Return %	35.44 %	35.44 %	N/A
All trades	23	23 (100.00 %)	0 (0.00 %)
Avg. Profit/Loss	45.34	45.34	N/A
Avg. Profit/Loss %	0.45 %	0.45 %	N/A
Avg. Bars Held	4.26	4.26	N/A
Winners	18 (78.26 %)	18 (78.26 %)	0 (0.00 %)
Total Profit	2500.86	2500.86	0.00
Avg. Profit	138.94	138.94	N/A
Avg. Profit %	1.39 %	1.39 %	N/A
Avg. Bars Held	3.17	3.17	N/A
Max. Consecutive	9	9	0
Largest win	442.64	442.64	0.00
# bars in largest win	3	3	0
Losers	5 (21.74 %)	5 (21.74 %)	0 (0.00 %)
Total Loss	-1457.97	-1457.97	0.00
Avg. Loss	-291.59	-291.59	N/A
Avg. Loss %	-2.92 %	-2.92 %	N/A
Avg. Bars Held	8.20	8.20	N/A
Max. Consecutive	4	4	0
Largest loss	-989.86	-989.86	0.00
# bars in largest loss	11	11	0
Max. trade drawdown	-1379.20	-1379.20	0.00
Max. trade % drawdown	-13.79 %	-13.79 %	0.00 %
Max. system drawdown	-1379.20	-1379.20	0.00
Max. system % drawdown	-1.35 %	-1.35 %	0.00 %
Recovery Factor	0.76	0.76	N/A
CAR/MaxDD	0.06	0.06	N/A
RAR/MaxDD	26.22	26.22	N/A
Profit Factor	1.72	1.72	N/A
Payoff Ratio	0.48	0.48	N/A
Standard Error	249.55	249.55	0.00
Risk-Reward Ratio	0.31	0.31	N/A
Ulcer Index	0.26	0.26	0.00
Ulcer Performance Index	-20.44	-20.44	N/A
Sharpe Ratio of trades	1.11	1.11	0.00
K-Ratio	0.0203	0.0203	N/A

Figure 9.9 -- Statistics for Long Aggressive Trades

Figure 9.10 shows a listing of the trades for the year 2011. The top section has the basic trades, the middle section has the aggressive trades, and the bottom section has all the trades.

Basic Trades

SPY	Long	2/24/2011 130.93	2/25/2011 132.33	1.07%
SPY	Long	3/16/2011 126.18	3/21/2011 129.74	2.82%
SPY	Long	5/5/2011 133.61	5/9/2011 134.72	0.83%
SPY	Long	6/6/2011 129.04	6/9/2011 129.4	0.28%
SPY	Long	7/12/2011 131.4	7/15/2011 131.69	0.22%
SPY	Long	7/27/2011 130.6	8/11/2011 117.33	-10.16%

Aggressive Trades

SPY	Long	6/7/2011 128.96	6/9/2011 129.4	0.34%
SPY	Long	7/14/2011 130.93	7/15/2011 131.69	0.58%
SPY	Long	7/28/2011 130.22	8/11/2011 117.33	-9.90%

Both Basic and Aggressive Trades

SPY	Long	2/24/2011 130.93	2/25/2011 132.33	1.07%
SPY	Long	3/16/2011 126.18	3/21/2011 129.74	2.82%
SPY	Long	5/5/2011 133.61	5/9/2011 134.72	0.83%
SPY	Long	6/6/2011 129.04	6/9/2011 129.4	0.28%
SPY	Long	6/7/2011 128.96	6/9/2011 129.4	0.34%
SPY	Long	7/12/2011 131.4	7/15/2011 131.69	0.22%
SPY	Long	7/14/2011 130.93	7/15/2011 131.69	0.58%
SPY	Long	7/27/2011 130.6	8/11/2011 117.33	-10.16%
SPY	Long	7/28/2011 130.22	8/11/2011 117.33	-9.90%

Figure 9.10 -- Trade list for 3 Day High Low

Figure 9.11 shows the plot with arrows at the entry and exit for both basic and aggressive trades. You can see the adjacent arrows in June, mid-July, and late-July where the aggressive trades were taken.

Figure 9.11 -- Plot showing all trades

COMMENTARY

One of the problems with mean reversion systems is that they are not self correcting. Several of the systems tested showed a trade that lost 10%. You can see such a trade in Figure 9.11. It is a long trade entered in late July and held until mid-August.

Most mean reversion systems can be designed to have both a basic entry and one or more aggressive entries. In many cases, the performance from the aggressive entries is better than from the basic entries. By programming so that each entry can be tested separately, and analyzed separately, you can decide which signals to include in your systems and which to avoid.

REGIME CHANGE

The regime change system has two sets of rules and parameters. Each set defines a model. They could be any models you want—trend following, mean reverting, pattern, whatever. In keeping with commonly used terminology, they are referred to as systems. They are arbitrarily named S2 and S3. As written, the two systems are both mean reverting and both are based on the RSI indicator. (There can be any number of systems—the code becomes a little more complicated, but the technique is the same.)

System S2 uses an RSI lookback length that is stored in the variable S2RSILB. This can be set or can be determined by optimization, including by the walk forward process. In the code shown in Listing 9.4, it is 9. S2 enters a long position when that RSI level drops below the value stored in S2BuyLevel—31 in the code below. It exits the long position when that RSI value rises above 75. The equity curve associated with S2 is computed and stored in S2Equity.

System S3 works the same way, using a length of S3RSILB, an entry at S3BuyLevel, storing the result in S3Equity.

There is no restriction on the systems defined by S2 and S3—all they need to do is generate buy and sell signals on their own.

At the close of every day, the recent performance of the systems is compared. That system whose equity has risen the most is identified as being *dominant*, and its signals are used as long as it continues to be dominant.

The looping code processes the data bar-by-bar. The overall system, RegimeChange, is in one of three states:
- State 0—flat.
- State 2—Long, from a signal from system S2.
- State 3—Long, from a signal from system S3.

How you want to handle changes in states and signals from the two systems is up to you. The code I have written does it this way:

> If there is a new dominant system, exit any open position unless there is a new buy signal from the new dominant system. If there is a continuing dominant system, follow its signals and ignore all others.

Systems

```
//   RegimeChange.afl
//
//   Keeps track of results from two sets of rules.
//   Uses whichever one has shown the best recent performance.
//
//   This is not a particularly good system.
//   It is intended to illustrate the concept and technique.
//

/////////////////////////////////////////
//   System Parameters
OptimizerSetEngine( "cmae" );
SetOption( "initialequity", 100000 );
MaxPos = 1;
SetOption( "MaxOpenPositions", MaxPos );
SetPositionSize( 10000, spsValue );
SetOption( "ExtraColumnsLocation", 1 );

SetTradeDelays( 0, 0, 0, 0 );
BuyPrice = SellPrice = Close;

ROCLB = Optimize( "ROCLB", 28, 2, 42, 1 );
S2BuyLevel = Optimize( "S2BuyLevel", 31, 1, 100, 1 );
S3BuyLevel = Optimize( "S3BuyLevel", 13, 1, 20, 1 );
S2RSILB = Optimize( "S2RSILB", 9, 2, 10, 1 );
S3RSILB = Optimize( "S3RSILB", 3, 2, 10, 1 );

//   System 2 -- Use RSI(2), or whatever length is working

RSI2 = RSI( S2RSILB );
Buy = S2Buy = RSI2 < S2BuyLevel;
Sell = S2Sell = RSI2 > 75;

S2Equity = Equity();
S2ROC = ROC( S2Equity, ROCLB );

//   System 3 -- Use RSI(3), or whatever other length is working

RSI3 = RSI( S3RSILB );
Buy = S3Buy = RSI3 < S3BuyLevel;
Sell = S3Sell = RSI3 > 75;

S3Equity = Equity();
S3ROC = ROC( S3Equity, ROCLB );

//   Choose which to use
//   Compare recent results

dominant = IIf( S2ROC > S3ROC, 2, 3 );

state[0] = 0;

for ( i = 1;i < BarCount;i++ )
{
//   If this bar is the first bar of a new state,
//   go flat unless there is a buy signal for the new state.
    if ( dominant[i] != dominant[i-1] )
    {
```

```
//   change in dominant system
     switch ( dominant[i] )
     {

        case 2:

            if ( S2Buy[i] )
            {
                Buy[i] = 1;
                state[i] = 2;
            }
            else
            {
                Sell[i] = 1;
                state[i] = 0;
            }    //  case 2

            break;

        case 3:
            if ( S3Buy[i] )
            {
                Buy[i] = 1;
                state[i] = 3;
            }
            else
            {
                Sell[i] = 1;
                state[i] = 0;
            }    // case 3

            break;
     }   //  switch
}   //  if
else
{
//   continuation of dominant system
     switch ( state[i-1] )
     {

         case 0:
             {
//  coming in flat
             if ( dominant[i] == 2 AND S2Buy[i] )
             {
                 Buy[i] = 1;
                 state[i] = 2;
             }
             else
                 if ( dominant[i] == 3 AND S3Buy[i] )
                 {
                     Buy[i] = 1;
                     state[i] = 3;
                 }
                 else
                 {
                     Buy[i] = 0;
                     state[i] = 0;
                 }
```

```
                    break;
                }   //   case 0

            case 2:
                {
//  long from system 2
                if ( dominant[i] == 2 AND S2Sell[i] )
                    {
                        Sell[i] = 1;
                        state[i] = 0;
                    }
                    else
                    {
                        state[i] = 2;
                    }

                    break;
                }   //   case 2

            case 3:
                {
//  long from system 3
                if ( dominant[i] == 3 AND S3Sell[i] )
                    {
                        Sell[i] = 1;
                        state[i] = 0;
                    }
                    else
                    {
                        state[i] = 3;
                    }

                    break;
                }   //   case 3

        }   //   switch
    }   //   else
}   //   for

e = Equity();

shapes = IIf( S2Buy, shapeUpArrow,
        IIf( S2Sell, shapeDownArrow, shapeNone ) );

shapecolors = IIf( S2Buy, colorGreen,
        IIf( S2Sell, colorRed, colorWhite ) );

Plot( C, "close", colorBlack, styleCandle );
PlotShapes( shapes, shapecolors );
Plot( dominant, "Dominant", colorGreen,
        styleLine | styleThick | styleOwnScale );

//Plot( state, "State", colorBlue, styleLine | styleOwnScale );
//Plot( e, "Equity", colorGreen, styleLine | styleOwnScale );

////////////////////   end   ////////////////////////
```

Listing 9.4 -- Regime Change System

Figure 9.12 shows a plot of the price along with a the dominant variable. When the heavy line is at the top, system S3 is dominant; when it is at the bottom, system S2 is dominant.

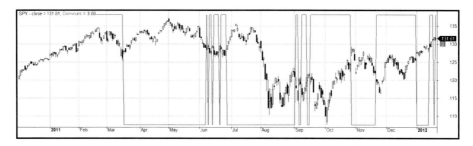

Figure 9.12 - Changing Dominant System

Figure 9.13 shows the equity curve trading SPY using the parameter values in the Listing 9.4.

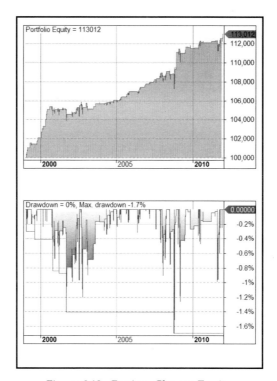

Figure 9.13 - Regime Change Equity

Figure 9.14 shows the statistics.

Statistics			
	All trades	Long trades	Short trades
Initial capital	100000.00	100000.00	100000.00
Ending capital	113012.02	113012.02	100000.00
Net Profit	13012.02	13012.02	0.00
Net Profit %	13.01 %	13.01 %	0.00 %
Exposure %	1.87 %	1.87 %	0.00 %
Net Risk Adjusted Return %	695.01 %	695.01 %	N/A
Annual Return %	0.95 %	0.95 %	0.00 %
Risk Adjusted Return %	50.52 %	50.52 %	N/A
All trades	84	84 (100.00 %)	0 (0.00 %)
Avg. Profit/Loss	154.91	154.91	N/A
Avg. Profit/Loss %	1.55 %	1.55 %	N/A
Avg. Bars Held	8.85	8.85	N/A
Winners	63 (75.00 %)	63 (75.00 %)	0 (0.00 %)
Total Profit	17410.07	17410.07	0.00
Avg. Profit	276.35	276.35	N/A
Avg. Profit %	2.76 %	2.76 %	N/A
Avg. Bars Held	8.03	8.03	N/A
Max. Consecutive	11	11	0
Largest win	1791.92	1791.92	0.00
# bars in largest win	5	5	0
Losers	21 (25.00 %)	21 (25.00 %)	0 (0.00 %)
Total Loss	-4398.05	-4398.05	0.00
Avg. Loss	-209.43	-209.43	N/A
Avg. Loss %	-2.09 %	-2.09 %	N/A
Avg. Bars Held	11.29	11.29	N/A
Max. Consecutive	2	2	0
Largest loss	-571.45	-571.45	0.00
# bars in largest loss	13	13	0
Max. trade drawdown	-1650.74	-1650.74	0.00
Max. trade % drawdown	-15.49 %	-15.49 %	0.00 %
Max. system drawdown	-1846.53	-1846.53	0.00
Max. system % drawdown	-1.69 %	-1.69 %	0.00 %
Recovery Factor	7.05	7.05	N/A
CAR/MaxDD	0.56	0.56	N/A
RAR/MaxDD	29.86	29.86	N/A
Profit Factor	3.96	3.96	N/A
Payoff Ratio	1.32	1.32	N/A
Standard Error	812.21	812.21	0.00
Risk-Reward Ratio	1.00	1.00	N/A
Ulcer Index	0.30	0.30	0.00
Ulcer Performance Index	-14.72	-14.72	N/A
Sharpe Ratio of trades	2.39	2.39	0.00
K-Ratio	0.0656	0.0656	N/A

Figure 9.14 - Regime Change Statistics

COMMENTARY

You might prefer to handle changes in state differently. Perhaps keep track of the shadow position of each system for all data bars. When a state change is detected, immediately take whatever position the new state is in. You will need additional variables and additional logic. I'll leave it as a homework problem.

Note that the 10% losing trade common in the other systems has been avoided.

This is designed to be a regime change system, not a voting system. One of the key components is determining which set of signals to use. This example uses a simple rate of change of equity. Alternatives might be based on statistical analysis of recent performance.

Connors RSI Pullback

This system is described in *Introduction to ConnorsRSI* by Larry Connors, Cesar Alvarez, and Matt Radtke. It is used here with permission.

The system has a lot of variables, rules, and parameters. It trades infrequently—too infrequently to be useful with SPY—but it does illustrate some interesting points.

1. Define new indicators. Streak Duration is defined as the number of consecutive closes in the same direction. Individual days are given a score of -1, 0, or +1 depending on whether the close is lower, equal to, or higher than the previous close. Form a series that increments each day the close is in the same direction. When the direction changes, reset to -1, 0, or +1.
2. Apply transformations to indicators. Compute the 2 period RSI of Streak Duration.
3. Combine indicators. Compute the average of three indicators that each are bounded in the range 0 to 100. The three are RSI of close, RSI of Streak Duration, and Percent Rank of one day price change. The result is the ConnorsRSI indicator.
4. Use a limit order to enter a long position at a significant discount from the previous close.

Listing 9.5 shows the AmiBroker code for the system.

```
//   ConnorsRSIPullback.afl
//
//   ConnorsRSI is defined in
//   "An Introduction to ConnorsRSI"
//   published by TradingMarkets.
//   It is used here with permission.
//
//   System settings

OptimizerSetEngine( "cmae" );

SetOption( "Initialequity", 100000 );
MaxPos = 1;
SetOption( "MaxOpenPositions", MaxPos );
SetPositionSize( 10000, spsValue );

SetOption( "ExtraColumnsLocation", 1 );
```

Systems

```
SetTradeDelays( 0, 0, 0, 0 );
BuyPrice = SellPrice = Close;

// User functions

function StreakDuration( p )
{
    t[0] = 0;

    for ( i = 1;i < BarCount;i++ )
    {
        t[i] = 0;

        if ( p[i] > p[i-1] )
            if ( t[i-1] >= 0 )
                t[i] = t[i-1] + 1;
            else
                t[i] = 1;
        else
            if ( p[i] < p[i-1] )
                if ( t[i-1] <= 0 )
                    t[i] = t[i-1] - 1;
                else
                    t[i] = -1;
    }

    return ( t );
}

function ConnorsRSI( V1, V2, V3 )
{
    t1 = RSIa( C, V1 );
    t2 = RSIa( StreakDuration( C ), V2 );
    OneDayReturn = ( C - Ref( C, -1 ) ) / Ref( C, -1 );
    t3 = PercentRank( OneDayReturn, V3 );
    return( ( t1 + t2 + t3 ) / 3 );
}

// Parameters

w = Optimize( "W", 8, 1, 20, 1 );     // Percent below previous close
x = Optimize( "x", 10, 5, 95, 5 );    // Position in Range of Close
y = Optimize( "y", 5, 5, 95, 5 );     // ConnorsRSI for entry
z = Optimize( "z", 10, 0, 16, 1 );    // Entry Limit
n = Optimize( "n", 80, 5, 95, 5 );    // ConnorsRSI for exit

// Indicators

LowBelowClose = Max( 100 * ( Ref( C, -1 ) - L ) / Ref( C, -1 ), 0 );
PIROfClose = 100 * ( C - L ) / ( H - L );
LimitEntry = ( 1 - 0.01 * z ) * C;

// Rules

ER1 = ADX( 10 ) > 30;
ER2 = LowBelowClose > w;
ER3 = PIROfClose < x;
ER4 = ConnorsRSI( 3, 2, 100 ) < y;
```

```
    ER5 = L < LimitEntry;

    XR1 = ConnorsRSI( 3, 2, 100 ) > n;

    //  Signals

    Buy = Ref( ER1, -1 )
          AND Ref( ER2, -1 )
          AND Ref( ER3, -1 )
          AND Ref( ER4, -1 )
          AND ER5;
    BuyPrice = LimitEntry;

    Sell = XR1;
    SellPrice = Close;

    //  Plots


    Plot( C, "C", colorBlack, styleCandle );


///////////////////  end  ///////////////////
```

Listing 9.5 -- Connors RSI Pullback System

COMMENTARY

When parameter values are set at the levels recommended by Connors, trading is very infrequent—less than one trade per issue per year, on average. The accuracy is very high, and the average gain per trade is very high.

This system has a very high number of "moving parts." In addition to the five variables that are exposed to searching, there are several more hard-coded into other indicators. There are five rules that must all be satisfied to enter a trade. Ordinarily, that many variables and rules would create a system that performs very well using the in-sample data, but fails out of sample. Connors RSI Pullback is very robust. It is profitable throughout its trading range.

Its profitability depends on it having many opportunities. The system returns very high profits during very concentrated periods. Trades come in bunches, separated by long periods of inactivity. Using a watchlist of current constituents of the S&P 100 index (be aware of survivorship bias), only 17 of the issues had any trades at all during the 13 year test period. Of five major ETFs (EEM, GLD, IWM, QQQ, and SPY) over 13 years, there was only a single trade—IWM in 2008.

Be prepared to wait months between trades for a given issue. The 17 trades using the S&P 100 watchlist came on 7 days—one each on two separate days in 1999, 3 on one day in 2001, 8 on one day in 2008, 2 on another day in 2008, 2 on one day in 2009, and 1 in 2011. The winning rate is high—100% of the 17 trades were winners. The percentage gained is spectacular—the average gain of the 17 trades was 30%, and the holding period was about two days.

Due to the very high correlation among trades when using a watchlist, the best estimate set of data used for the simulation to determine risk, position size, and profit potential cannot use trades. Instead, use changes in daily equity. When you use daily data, you will mark your equity to market every day, exposing intra-trade drawdown. Read Chapter 10 for a discussion of the implications.

Dual Time Frame

The concept behind the dual time frame system is to enter long trades when price is oversold on a short term basis, but not oversold on a longer term basis.

The system computes two indicators:
- RSI of daily bars.
- RSI of weekly bars.

To enter a long trade, require weekly RSI to be higher than some limit, suggesting that the issue is not oversold on a weekly basis, and daily RSI to be lower than some other limit, suggesting that it is oversold on a daily basis.

The basic RSI system enters a long position when the RSI using a short lookback length, say 2 or 3 days, falls below a low level, say 23. The rule is:

```
Buy = RSI(2) < 23;
```

Listing 9.6 shows the AmiBroker code using only daily RSI.

```
//   DualTimeFrame-DailyComponent.afl
//
//   Take positions when RSI on daily data is
//   at a low level

//   System settings

OptimizerSetEngine( "cmae" );

SetOption( "Initialequity", 100000 );
MaxPos = 1;
SetOption( "MaxOpenPositions", MaxPos );
SetPositionSize( 10000, spsValue );

SetOption( "ExtraColumnsLocation", 1 );
```

```
SetTradeDelays( 0, 0, 0, 0 );
BuyPrice = SellPrice = Close;

//   Parameters

//   2  23      CAR/MDD
//   2  8       K-ratio

DailyRSILB = Optimize( "DailyRSILB", 2, 2, 6, 1 );
DailyBuyLevel = Optimize( "DailyBuyLevel", 23, 5, 30, 1 );

//   Indicators

DailyRSI = RSI( DailyRSILB );

//   Rules

//   Signals

Buy = DailyRSI < DailyBuyLevel;
Sell = DailyRSI > 75;

//   Plots

Plot( C, "Close", colorBlack, styleCandle );

shapes = IIf( Buy, shapeUpArrow, shapeNone );
shapecolors = IIf( Buy, colorGreen, colorWhite );
PlotShapes( shapes, shapecolors );

Plot( DailyRSI, "DailyRSI", colorBlue,
        styleLine | styleThick | styleLeftAxisScale );

//////////////////// end ////////////////////
```

Listing 9.6 -- Dual Time Frame System—Daily Component

There are two parameters—the length of the RSI and the level signifying oversold. The best values are dependent on the objective function used to rank the alternatives. Figure 9.15 shows the equity curve when the values are 2 and 23, respectively, chosen when ranked by CAR/MDD.

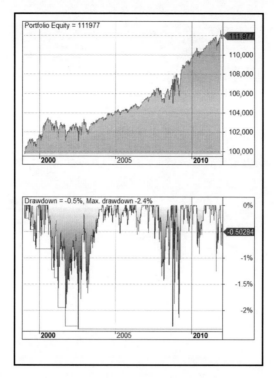

Figure 9.15 - Equity Curve - Daily RSI

Figure 9.16 shows the statistics.

	All trades	Long trades	Short trades
Statistics			
Initial capital	100000.00	100000.00	100000.00
Ending capital	111976.68	111976.68	100000.00
Net Profit	11976.68	11976.68	0.00
Net Profit %	11.98 %	11.98 %	0.00 %
Exposure %	4.19 %	4.19 %	0.00 %
Net Risk Adjusted Return %	286.01 %	286.01 %	N/A
Annual Return %	0.87 %	0.87 %	0.00 %
Risk Adjusted Return %	20.88 %	20.88 %	N/A
All trades	239	239 (100.00 %)	0 (0.00 %)
Avg. Profit/Loss	50.11	50.11	N/A
Avg. Profit/Loss %	0.50 %	0.50 %	N/A
Avg. Bars Held	7.08	7.08	N/A
Winners	167 (69.87 %)	167 (69.87 %)	0 (0.00 %)
Total Profit	29798.19	29798.19	0.00
Avg. Profit	178.43	178.43	N/A
Avg. Profit %	1.78 %	1.78 %	N/A
Avg. Bars Held	4.97	4.97	N/A
Max. Consecutive	10	10	0
Largest win	1082.69	1082.69	0.00
# bars in largest win	6	6	0
Losers	72 (30.13 %)	72 (30.13 %)	0 (0.00 %)
Total Loss	-17821.50	-17821.50	0.00
Avg. Loss	-247.52	-247.52	N/A
Avg. Loss %	-2.48 %	-2.48 %	N/A
Avg. Bars Held	11.99	11.99	N/A
Max. Consecutive	3	3	0
Largest loss	-1316.33	-1316.33	0.00
# bars in largest loss	20	20	0
Max. trade drawdown	-2474.41	-2474.41	0.00
Max. trade % drawdown	-23.75 %	-23.75 %	0.00 %
Max. system drawdown	-2474.41	-2474.41	0.00
Max. system % drawdown	-2.36 %	-2.36 %	0.00 %
Recovery Factor	4.84	4.84	N/A
CAR/MaxDD	0.37	0.37	N/A
RAR/MaxDD	8.86	8.86	N/A
Profit Factor	1.67	1.67	N/A
Payoff Ratio	0.72	0.72	N/A
Standard Error	1041.06	1041.06	0.00
Risk-Reward Ratio	0.77	0.77	N/A
Ulcer Index	0.58	0.58	0.00
Ulcer Performance Index	-7.82	-7.82	N/A
Sharpe Ratio of trades	0.77	0.77	0.00
K-Ratio	0.0504	0.0504	N/A

Figure 9.16 - Statistics - Daily RSI

Using AmiBroker's TimeFrame functions, daily data is consolidated into weekly, which are then used to compute an indicator. The level of that indicator is used as a permission filter to allow or block trades. Listing 9.7 shows the AmiBroker code.

```
//   DualTimeFrame.afl
//
//   Use RSI on weekly data as a filter
//   Take positions when RSI on daily data is
//   at a low level
```

```
//   System settings

OptimizerSetEngine( "cmae" );

SetOption( "Initialequity", 100000 );
MaxPos = 1;
SetOption( "MaxOpenPositions", MaxPos );
SetPositionSize( 10000, spsValue );

SetOption( "ExtraColumnsLocation", 1 );

SetTradeDelays( 0, 0, 0, 0 );
BuyPrice = SellPrice = Close;

//   Parameters

//   3  15  6  46     HBMetric 56 trades
//   3  15  5  64  K-ratio     23 trades
//   2  25  5  56  Best CAR/MDD with >100 trades

DailyRSILB = Optimize( "DailyRSILB", 3, 2, 6, 1 );
DailyBuyLevel = Optimize( "DailyBuyLevel", 15, 5, 30, 1 );
WeeklyRSILB = Optimize( "WeeklyRSILB", 5, 1, 10, 1 );
WeeklyPermitLevel = Optimize( "WeeklyPermitLevel", 64, 10, 90, 1 );

//   Indicators

//   The data base is daily.
//   Form weekly bars.

TimeFrameSet( inWeekly );
WRSI = RSI( WeeklyRSILB );
TimeFrameRestore();

//   Expand weekly indicator to daily periodicity
WeeklyRSI = TimeFrameExpand( WRSI, inWeekly );
DailyRSI = RSI( DailyRSILB );

//   Rules

Permission = WeeklyRSI > WeeklyPermitLevel;

//   Signals

Buy = Permission AND DailyRSI < DailyBuyLevel;
Sell = DailyRSI > 75;  //  OR !Permission;
Buy = ExRem( Buy, Sell ); //   remove extraneous arrows
Sell = ExRem( Sell, Buy );

//   Plots

Plot( C, "Close", colorBlack, styleCandle );

shapes = IIf( Buy, shapeUpArrow,
         IIf( Sell, shapeDownArrow, shapeNone ) );
shapecolors = IIf( Buy, colorGreen,
         IIf( Sell, colorRed, colorWhite ) );
PlotShapes( shapes, shapecolors );
```

```
Plot( WeeklyRSI, "WeeklyRSI", colorGreen,
      styleDots | styleThick | styleLeftAxisScale );
Plot( DailyRSI, "DailyRSI", colorBlue,
      styleLine | styleThick | styleLeftAxisScale );
```

/////////////////////// end ///////////////////////

Listing 9.7 - Dual Time Frame System—Both Components

The four parameter values are 3, 15, 5, 64. The parameters that indicated oversold using daily data moved from 2 and 23 to 3 and 15. The weekly RSI, based on a 5 week lookback, must be above 64 to permit long trades. Figure 9.17 shows the equity curve.

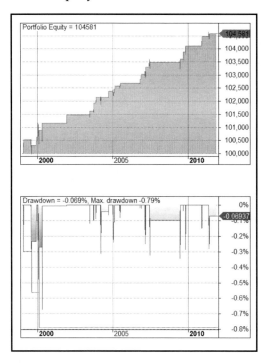

Figure 9.17 - Equity Curve - Dual

Figure 9.18 shows the statistics.

Statistics	All trades	Long trades	Short trades
Initial capital	100000.00	100000.00	100000.00
Ending capital	104580.70	104580.70	100000.00
Net Profit	4580.70	4580.70	0.00
Net Profit %	4.58 %	4.58 %	0.00 %
Exposure %	0.65 %	0.65 %	0.00 %
Net Risk Adjusted Return %	704.02 %	704.02 %	N/A
Annual Return %	0.35 %	0.35 %	0.00 %
Risk Adjusted Return %	53.07 %	53.07 %	N/A
All trades	23	23 (100.00 %)	0 (0.00 %)
Avg. Profit/Loss	199.16	199.16	N/A
Avg. Profit/Loss %	1.99 %	1.99 %	N/A
Avg. Bars Held	10.43	10.43	N/A
Winners	21 (91.30 %)	21 (91.30 %)	0 (0.00 %)
Total Profit	4840.10	4840.10	0.00
Avg. Profit	230.48	230.48	N/A
Avg. Profit %	2.30 %	2.30 %	N/A
Avg. Bars Held	9.71	9.71	N/A
Max. Consecutive	11	11	0
Largest win	525.78	525.78	0.00
# bars in largest win	11	11	0
Losers	2 (8.70 %)	2 (8.70 %)	0 (0.00 %)
Total Loss	-259.40	-259.40	0.00
Avg. Loss	-129.70	-129.70	N/A
Avg. Loss %	-1.30 %	-1.30 %	N/A
Avg. Bars Held	18.00	18.00	N/A
Max. Consecutive	1	1	0
Largest loss	-211.40	-211.40	0.00
# bars in largest loss	17	17	0
Max. trade drawdown	-796.06	-796.06	0.00
Max. trade % drawdown	-7.57 %	-7.57 %	0.00 %
Max. system drawdown	-796.06	-796.06	0.00
Max. system % drawdown	-0.79 %	-0.79 %	0.00 %
Recovery Factor	5.75	5.75	N/A
CAR/MaxDD	0.44	0.44	N/A
RAR/MaxDD	67.43	67.43	N/A
Profit Factor	18.66	18.66	N/A
Payoff Ratio	1.78	1.78	N/A
Standard Error	160.30	160.30	0.00
Risk-Reward Ratio	2.06	2.06	N/A
Ulcer Index	0.08	0.08	0.00
Ulcer Performance Index	-63.10	-63.10	N/A
Sharpe Ratio of trades	6.07	6.07	0.00
K-Ratio	0.1351	0.1351	N/A

Figure 9.18 - Statistics - Dual Time Frame

Figure 9.19 shows a plot of the price of SPY, along with the two RSI indicators, and the buy and sell arrows for three trades.

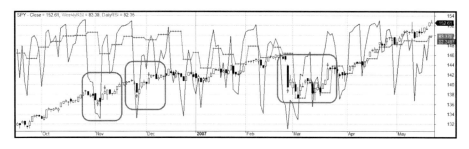

Figure 9.19 - Plot of Dual Time Frame RSIs

COMMENTARY

There are several system design options that you can use.
- The example uses weekly data for the permission filter. Weekly bars always close on the final trading day of the week—usually Friday. AmiBroker allows an almost-equivalent n-day bar. When n is 5, the 5-day bars are almost weekly bars—they always include 5 days (some weekly bars have fewer than 5 days). 5-day bars do not always close on Friday. In fact, changing the starting date of a test changes the sequence of ending days for the 5-day bars and may change the test results. If results do change significantly, that is an indication that the system is not robust.
- This example tests the permission filter when entering a new trade, but not when exiting an existing trade. The logic could be written so that an exit is signaled by either the daily RSI or the weekly RSI. The sell statement would be:

    Sell = RSI(2) > 75 OR !Permission;

    You can test that easily by removing the comment marks from the Sell statement in the example code.

Using any permission filter reduces the number of trades. If the large losing trades are avoided, that is a benefit. But reducing the number of trades also reduces the number of compounding opportunities.

Avoiding losing trades is important, particularly large losses. In this example, there is a trade that loses 13% when only daily RSI is used. But the worst trade for the dual time frame system is 2%.

Comparing the statistics, adding the weekly RSI as a filter changed the performance characteristics. The average gain per trade increased from about 0.5% to about 2.0%, while the average holding period increased from about 7 days to about 10 days, with some of the trades held about 20 trading days. That is no longer the swing trading system we originally had in mind.

Whether performance improves may not be immediately apparent.

To fully understand the value of this, or any, permission filter, work through the estimation of risk, position size, and profit potential as described in *Modeling Trading System Performance*. Normalize the two alternatives—filtered and unfiltered—at a common risk, compute the position size for each, then compare the distribution of profit potential.

VIX

VIX is a data series derived from the volatility of the S&P 500 futures contract. Simplified, VIX is the standard deviation of price changes. VIX seldom trends—it is very mean reverting. And it is inversely correlated with SPY, as Figure 9.20 shows.

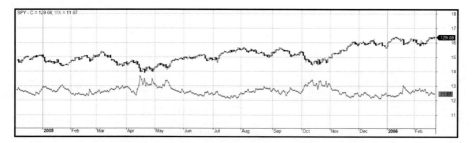

Figure 9.20 - SPY and VIX

Figure 9.21 shows the scatter plot of 1 day change in SPY related to 1 day change in VIX. There are 3271 data points for the 13 year test period. VIX is about 6 times as volatile as SPY.

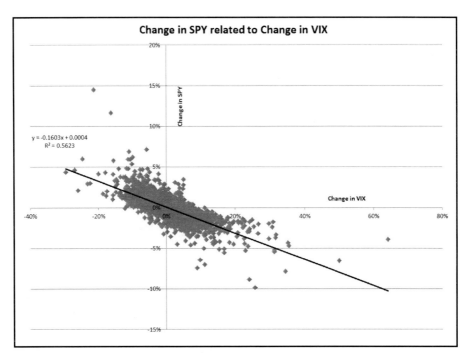

Figure 9.21 - Change in SPY Related to Change in VIX

Earlier, we had applied a zigzag indicator to the SPY price series and adjusted the percentage until we found the change that produced about 24 bottoms per year. We found that it ranged between 0.7% and 2.1% over the 13 year test period. Using that same technique on the VIX price series, we find that a zigzag with changes of about 7% result in about 24 bottoms per year, as shown in Figure 9.22.

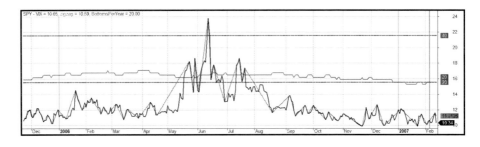

Figure 9.22 - VIX with Count of Bottoms per Year

Rises in VIX are associated with drops in SPY and vice versa. A model that identifies tops and bottoms in VIX might be useful in generating signals to trade SPY or some other issue. Let's test that first. Listing 9.8 is a program that uses perfect entries based on VIX to trade another issue.

```
//    FindSurrogate.afl
//
//    Identify perfect tops and bottoms for VIX
//    based on the ZigZag function with 7% change.
//
//    Take a long position in the issue being tested
//    for use as a surrogate at every VIX top
//    and exit at the VIX bottom.

//  System settings

OptimizerSetEngine( "cmae" );

SetOption( "Initialequity", 100000 );
MaxPos = 1;
SetOption( "MaxOpenPositions", MaxPos );
SetPositionSize( 10000, spsValue );

SetOption( "ExtraColumnsLocation", 1 );

SetTradeDelays( 0, 0, 0, 0 );
BuyPrice = SellPrice = Close;

//   Parameters

SetForeign( "$VIX" );

zz = Zig( C, 7 );

bottom = zz < Ref( zz, 1 ) AND zz < Ref( zz, -1 );
top = zz > Ref( zz, 1 ) AND zz > Ref( zz, -1 );
RestorePriceArrays();

Buy = top;
Sell = bottom;

/////////////////////   end   /////////////////////////
```

Listing 9.8 - Find Surrogates

Mean Reversion Trading Systems

The program was tested on the list of issues mentioned at the beginning of the book—major ETFs and a variety of issues that are net flat over the 13 year test period. Figure 9.23 shows some summary results.

Ticker	Net Profit	K-Ratio	# Trades	Avg % Profit/Loss	Avg Bars Held	% of Winners
MS	211,037.84	0.0766	317	6.66	6.71	87.70
COF	193,766.55	0.0794	317	6.11	6.71	82.02
BAC	156,053.90	0.0473	317	4.92	6.71	84.54
CSCO	148,507.54	0.1123	317	4.68	6.71	84.54
TWX	144,303.08	0.0868	317	4.55	6.71	82.33
INTC	139,860.12	0.1378	317	4.41	6.71	82.65
F	137,339.08	0.0635	317	4.33	6.71	75.39
XLF	135,166.34	0.0725	317	4.26	6.71	91.80
QQQ	134,480.93	0.1273	311	4.32	6.76	94.86
BK	132,899.80	0.0917	317	4.19	6.71	81.39
SMH	129,501.34	0.1270	281	4.61	6.85	84.34
CMCSA	126,908.43	0.0934	317	4.00	6.71	80.44
XLK	125,704.22	0.1290	317	3.97	6.71	94.64
DELL	124,426.56	0.0994	317	3.93	6.71	79.18
HD	124,042.00	0.1260	317	3.91	6.71	82.65
GE	121,464.24	0.0864	317	3.83	6.71	84.23
WY	114,176.54	0.0780	317	3.60	6.71	84.23
XLB	112,654.66	0.0970	317	3.55	6.71	88.01
IWM	112,186.61	0.1124	279	4.02	6.84	95.70
XLY	111,177.94	0.1238	317	3.51	6.71	92.43
EEM	110,731.50	0.0426	218	5.08	6.59	94.50
DIS	110,375.16	0.1100	317	3.48	6.71	79.18
XLE	108,841.34	0.0747	317	3.43	6.71	87.07
XLI	107,559.61	0.1214	317	3.39	6.71	93.38
DD	104,857.66	0.0786	317	3.31	6.71	81.39
SPY	104,702.95	0.1249	317	3.30	6.71	97.16
MSFT	104,213.17	0.1305	317	3.29	6.71	78.55
ALL	94,042.75	0.0588	317	2.97	6.71	78.23
IVV	90,943.64	0.1080	280	3.25	6.85	96.79
PFE	82,611.30	0.0862	317	2.61	6.71	74.13
EWJ	79,429.86	0.1127	317	2.51	6.71	82.65
XLV	78,773.41	0.1261	317	2.48	6.71	91.80
T	74,670.62	0.1035	317	2.36	6.71	78.23
RTN	69,065.75	0.1513	317	2.18	6.71	74.45
VZ	68,738.38	0.0902	317	2.17	6.71	73.82
XLU	65,225.54	0.1144	317	2.06	6.71	85.17
MRK	65,189.65	0.0788	317	2.06	6.71	73.82
AEP	63,773.09	0.1256	317	2.01	6.71	75.39
XLP	56,118.17	0.1218	317	1.77	6.71	83.91
KO	49,459.20	0.0824	317	1.56	6.71	74.45
HNZ	47,041.91	0.0652	317	1.48	6.71	76.97
GLD	16,603.75	0.0330	187	0.89	6.37	64.71

Figure 9.23 - Surrogates for VIX - Summary Results

Every one of the potential surrogates was profitable. We expect all 317 of the VIX positions to be profitable, because we used known top and bottom points of the zigzag. The %Winners column shows that about 80 to 90 percent of surrogate traders were winners. Several highly liquid equities ranked at the top of the list, and all of the major ETFs, with the exception of GLD, show excellent results.

Next, we need a system that trades VIX accurately. Listing 9.9 is a simple RSI system. Some parameter sets work well, others leave an equity curve that is ragged.

```
//    TradeSPYFromVIX.afl
//
//    Use VIX as the source of data
//    for the signals.
//    When VIX is high, buy the surrogate.
//
//    This system uses a 2 period RSI of VIX.
//

//  System settings

OptimizerSetEngine( "cmae" );

SetOption( "Initialequity", 100000 );
MaxPos = 1;
SetOption( "MaxOpenPositions", MaxPos );
SetPositionSize( 10000, spsValue );

SetOption( "ExtraColumnsLocation", 1 );

SetTradeDelays( 0, 0, 0, 0 );
BuyPrice = SellPrice = Close;

//   parameters

RSILB = Optimize( "RSILB", 4, 2, 6, 1 );
BuyLevel = Optimize( "BuyLevel", 83, 1, 99, 1 );
SellLevel = Optimize( "SellLevel", 9, 1, 99, 1 );
ProfitTarget = Optimize( "ProfitTarget", 0.5, 0.5, 2, 0.25 );

//   Indicators

VIXC = Foreign( "$VIX", "C" );

RSI2 = RSIa( VIXC, RSILB );

Buy = RSI2 > BuyLevel;
Sell = RSI2 < SellLevel;

ApplyStop( stopTypeProfit, stopModePercent, profittarget );

////////////   end   ////////////////////
```

Listing 9.9 - Trade SPY From VIX Signals

Figure 9.24 shows the equity curve.

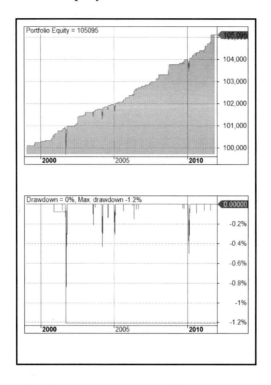

Figure 9.24 - Equity Curve Trading SPY from Signals from VIX

Figure 9.25 shows the statistics.

Statistics	All trades	Long trades	Short trades
Initial capital	100000.00	100000.00	100000.00
Ending capital	105094.51	105094.51	100000.00
Net Profit	5094.51	5094.51	0.00
Net Profit %	5.09 %	5.09 %	0.00 %
Exposure %	0.66 %	0.66 %	0.00 %
Net Risk Adjusted Return %	769.63 %	769.63 %	N/A
Annual Return %	0.38 %	0.38 %	0.00 %
Risk Adjusted Return %	57.88 %	57.88 %	N/A
All trades	80	80 (100.00 %)	0 (0.00 %)
Avg. Profit/Loss	63.68	63.68	N/A
Avg. Profit/Loss %	0.64 %	0.64 %	N/A
Avg. Bars Held	3.79	3.79	N/A
Winners	80 (100.00 %)	80 (100.00 %)	0 (0.00 %)
Total Profit	5094.51	5094.51	0.00
Avg. Profit	63.68	63.68	N/A
Avg. Profit %	0.64 %	0.64 %	N/A
Avg. Bars Held	3.79	3.79	N/A
Max. Consecutive	80	80	0
Largest win	209.81	209.81	0.00
# bars in largest win	2	2	0
Losers	0 (0.00 %)	0 (0.00 %)	0 (0.00 %)
Total Loss	0.00	0.00	0.00
Avg. Loss	N/A	N/A	N/A
Avg. Loss %	N/A	N/A	N/A
Avg. Bars Held	N/A	N/A	N/A
Max. Consecutive	0	0	0
Largest loss	0.00	0.00	0.00
# bars in largest loss	0	0	0
Max. trade drawdown	-1217.84	-1217.84	0.00
Max. trade % drawdown	-12.18 %	-12.18 %	0.00 %
Max. system drawdown	-1217.84	-1217.84	0.00
Max. system % drawdown	-1.21 %	-1.21 %	0.00 %
Recovery Factor	4.18	4.18	N/A
CAR/MaxDD	0.32	0.32	N/A
RAR/MaxDD	47.96	47.96	N/A
Profit Factor	N/A	N/A	N/A
Payoff Ratio	N/A	N/A	N/A
Standard Error	148.98	148.98	0.00
Risk-Reward Ratio	2.55	2.55	N/A
Ulcer Index	0.06	0.06	0.00
Ulcer Performance Index	-81.04	-81.04	N/A
Sharpe Ratio of trades	16.82	16.82	0.00
K-Ratio	0.1674	0.1674	N/A
HBMetric	14702.59		

Figure 9.25 - Statistics Trading SPY from Signals from VIX

Summary

This system takes advantage of the characteristics of VIX as very mean reverting, and the correlation between SPY and VIX. Your research into other systems that could be used to replace the simple system based on RSI will be rewarded.

CHAPTER SUMMARY

We have demonstrated use of several different indicators and techniques for generating buy and sell signals to trade SPY. Each of the systems shown is profitable in the backtests reported. While the systems appear to be stable and robust, backtest results are not sufficient, no matter how good they look. Before trading one of these systems, or any system, you must perform your own testing, validation, and analysis. You must understand how the system works, the risk of drawdown, the maximum safe position size, and the profit potential. You must have a set of trades and performance metrics to use as a baseline, and to which you will compare actual performance.

10

Multi-System Systems

To set the stage for this chapter, let me reiterate my recommendation for system characteristics:
- Trades a single highly liquid issue, say SPY.
- Is either long or flat. (There might be another system that is short or flat.)
- Trades frequently, say 24 times per year.
- Holds for a short period, say 1 to 3 days.
- Has a high percentage of winning trades. 65% is a reasonable and achievable minimum.
- Has a positive expectancy.
- Limits losses on losing trades.
- Passes tests of validation on out-of-sample data. Walk forward is best.

It is difficult to meet all of these criteria in a single system. One tactic to overcome this is to combine signals from several systems, each of which is a cherry picker — it seeks the best, most reliable, most profitable, least risky patterns and trades infrequently.

There are three major areas of concern related to the individual systems and the system of systems that result from combining them:
- However the individual systems, or the system of systems, are defined and constructed, they must all pass validation.
- In total, there are a large number of parameters and a risk of over-fitting.
- There will very likely be positive correlation among the trades.

Validation

You need confidence your system will perform well when using real money. All development introduces biases and distortions. The least biased estimate of future performance comes from the out-of-sample trades of the walk forward process. The estimates of risk and profit potential come from Monte Carlo simulations for the time horizon you plan to trade, say two years.

The ideal, arrived at by maximizing exposure when conditions are best while remaining flat when there is not a well defined profit potential, is to have long positions 30 to 40 percent of the time, short positions 30 to 40 percent of the time, and flat 20 to 40 percent of the time. For the system that takes long positions, 30% of a 252 day trading year is 75 days. This could be any combination of 15 5-day trades to 37 2-day trades. 24 3-day trades is a reasonable goal and fits the concept of two cycles each month. 72 days is 28% of a trading year. A system that trades 24 times a year has 48 trades in the two year forecast.

A cherry picking system trades less frequently. It might have a position 5% of the time — 12 days a year, coming from 4 3-day trades. Together, six such systems have 24 3-day trades per year.

Difficulties arise during validation and analysis. Taken individually, with only 4 trades per year there will be long periods of inactivity. The two year forecast has only 8 trades on average. Given a period of poor performance, with so few trades, it will difficult to determine whether the system is performing within expectations or is broken. The distribution of risk and profit will show wide ranges and high uncertainty. In order to keep risk low, position size will be low. Profit potential in correspondingly low.

Degrees of Freedom

Six cherry picking systems, all trading SPY, could be, and probably should be coded together to coordinate signals.

Assume each of the six systems has four subjective decisions and four variable parameters. Together they have 24 subjective decisions and 24 variable parameters in order to achieve 24 trades per year. With that many moving parts, it will be difficult to isolate the problem when drawdowns occur, and tempting to tweak one of the many variables to correct the performance.

Seen as a single system, entering validation as a single system, the large number of parameters that must be selected causes a serious problem of the *curse of dimensionality*. For more discussion, see Chapter 2.

POSITIVE CORRELATION

Ideally, the six individual systems each contributing four 3-day trades per year would have combined positions that fell on 72 separate days, with each 3-day trade independent and separate from all the others. It would take clever system design to have no overlap. Recall the coincident birthday problem, where there is a 50% probability that among any group of at least 23 people at least two people will share a birthday. If the days of the trades are selected at random, the probability of at least one duplication in any one of the 252 days is 50% when there are 18 days. Since all six systems are looking for the same cherries, the distribution of the 72 days is definitely not random, and the overlap is certain.

Positive correlation definitely increases risk. A simple thought experiment shows why.

Assume there are two systems, each with a positive expectancy, each producing 25 trades in a given year, each of which lasts exactly one day. If none of the trades occur on the same days, the correlation will be negative. If it is just win or loss that is being recorded for each day, the correlation will be about -0.10. A Monte Carlo simulation of the combined systems will select a day at random and record the trade from each that occurred on that day. On the day there is a trade from one system, there is never a trade from the other. The 50 trades will be spread among 50 different days. If there is a large losing trade on any given day, its effect is diluted by the other trades of the year. We get the benefit of compounding the additional trades along with the benefit of lowering risk. If both systems have mean geometric return per trade of 1%, each systems returns 28.2% per year, compound rate of return — 1.01 ^ 25 = 1.282. Combined, the two systems return 64.4% per year — 1.01 ^ 50 = 1.644.

Contrast with those same two systems but have them be perfectly correlated—the trades from both systems all occur on exactly the same days. On a day when one system has a losing trade, the other also has a losing trade. The effect on account equity is a drawdown twice as large as either individual system. Certainly the two uncorrelated systems could have losing trades on successive days, and that would total the same larger drawdown, but the probability of that happening is low. When the two systems are correlated, the probability is certain. The CAR is also a little lower at 64.0% — 1.02 ^ 25 = 1.640 — because you do not have the gain from the first trade of the day to help fund the second trade of the day. Note also that you may not be able to fund all of the signaled trades when they are correlated. If the position size is 60% of the account for each system, and you cannot use margin or leverage, two signals on the same day that call for 120% cannot be fully funded.

Since it is the size and number of losing trades that limit the position size, the safe position size for the correlated systems will be smaller than for the uncorrelated systems. Even if the profit potential of the two alternatives is the same, the smaller position size results in lower profit for the correlated systems at the same level of risk.

Using Equity Changes for Analysis

Because of the difficulty aligning trades to days and determining the degree of overlap, it is easier to use daily equity changes than to use closed trades when performing the simulations.

Note that using equity changes will expose a greater portion of the intra-trade drawdowns. Assume a trade lasts three days and the returns on the days are +1%, -2%, +3%. The net trade is +2%.

- If the simulations are using closed trades, there is a single entry for +2%. The 2% drawdown in the trade is seen as a maximum adverse excursion, but not as a drawdown. (Depending on the development platform, it may not even be seen as an MAE in its entirety. Some platforms measure MAE from entry, others from maximum equity. If the measurement is from entry, this trade would show an MAE of 1%.)
- If the simulations are using daily equity changes, there are three entries of +1%, -2%, and +3%, and the 2% loss is seen as a drawdown. If you are managing your trades such that you will exit a trade based on open trade drawdown, then you should recognize the 2% loss immediately. Your simulations should be performed using daily equity changes rather than closed trades.

If you would exit from an open trade based on an intra-day drawdown, but prior to the exit signaled by the model itself, then your analysis must be done in still more detail, taking daily high and low prices into account.

You always *can* use daily equity changes to study the risk and profit potential. There are three situations where, in order to get accurate results, you *must* use daily equity changes:

- Whenever the system is a combination of other systems. Any time you cannot be certain whether there could be overlapping trades,

assume that there could be.
- Whenever positions might be added, such as scaling-in or taking additional trades. That is, responding to additional signals that occur when there is already a trade open and taking additional positions. For mean reversion systems, the additional signals are often stronger and have better performance, and you probably do want to take the additional positions. But, be certain you understand the increased risk.
- When you would exit a trade to take the system offline due to excessive drawdown based on intra-trade drawdown—compared with waiting until the trade is closed and recalculating the position size prior to entering the next trade.

Summary

With multi-system systems you have two choices, each with its own set of complications and considerations.

Combine the logic from the separate systems into a single system.

You must:
- Validate and analyze it as a single system.
- Deal with the high number of parameters and likelihood of overfitting.
- Avoid the temptation to fiddle with individual parameters.

You gain:
- Having a single system that trades often enough to provide a best estimate set of results and estimates of system performance.
- More accurate determination of position size based on the combination of signals.
- Earlier warning of underperformance of the entire multi-system.

Run several individual systems.
You must:
- Validate and analyze each system separately.
- Understand that infrequent trading makes validation of each system less certain.
- Understand that the number of months required to accumulate a meaningful trade history makes determination of individual system health difficult.
- Understand that there will be positive correlation. Systems that seek the same entry points are correlated, even if their methods

are different. Signals are not random; they are clustered around high probability patterns.
- Understand that, in periods of crisis, correlation increases. Correlation results in overestimation of profit and underestimation of risk.
- Understand that multiple signals may result in position size exceeding ability to fund all positions.

11

It Doesn't Work Anymore

The law that entropy always increases holds, I think, the supreme position among the laws of Nature. If someone points out to you that your pet theory of the universe is in disagreement with Maxwell's equations—then so much the worse for Maxwell's equations. If it is found to be contradicted by observation—well, these experimentalists do bungle things sometimes. But if your theory is found to be against the second law of thermodynamics I can give you no hope; there is nothing for it but to collapse in deepest humiliation.

—Sir Arthur Stanley Eddington, 1927

While writing, I received a telephone call from a consulting client. His firm was trading a mean reversion system that was well known and regularly discussed in print and in forums. Performance was excellent through 2011, then declined in 2012. We talked about why the system was failing and what to do about it.

Systems stop working when the model no longer fits the data; when the signals no longer precede profitable trades; when profitable trades no longer exist.

All trading systems, regardless of the method used for entry or the length of the holding period, are profitable only when they can identify a pattern that indicates an inefficiency in the price—the signal within the data. By making a profitable trade, a portion of that inefficiency is removed, creating a more nearly efficient market. As more traders recognize and act on the same inefficiency, sharing a finite pool of potential profit, each trade returns a smaller profit. Eventually those profits disappear and the equity curve associated with that system flattens out.

There is no doubt that wider knowledge and applications of the rules of a trading system will cause that system to be less profitable for everyone who trades it. But it does not take widespread knowledge of a specific system for its profitability to decline. Every profitable system is recognizing and exploiting some inefficiency. All it takes for a system to lose its edge is for that inefficiency to disappear.

The larger the inefficiency, the easier it is to recognize it and the more profitable it is. Do not expect to find large inefficiencies—they are gone, like the old-growth forests of the Northwest. The first loggers to find them harvested them, and that opportunity will never occur again. Smaller inefficiencies are tradable. This book is devoted to helping find them. Once they are gone, there will be only noise, and success will depend only on luck.

The Donchian breakout system, for example, has lost its edge. In part because more traders know its rules and are applying them. But in larger part because more traders are using more trend following systems and the inefficiencies that they all need are disappearing. In this, trading resembles thermodynamics. Markets evolve from inefficient to efficient. That process is irreversible. All systems will eventually degrade to the point where profit equals execution cost. There will continue to be winners and losers, but the reason will increasingly be due to luck rather than having a sustainable edge.

My expectation is that inefficiencies will continue to disappear, progressing from large profits recognized using monthly or weekly data downward to smaller profits found using well publicized technical indicators applied to end-of-day daily data and continuing to finer resolution price bars with more sophisticated methods.

To restore profitability if, or perhaps the better term is when, your system stops working:
- Rerun the validation — probably the walk forward runs. It will bring the logic and parameter values of the model back into synchronization with the data. If the out-of-sample results, particularly those of the most recent period, are profitable, the system can be returned to trading.
- Redesign the system to use finer resolution data. If you had been using only closing prices for daily bars, consider using daily open,

high, and low. Or perhaps prices for shorter time periods, say hourly bars.
- Redesign the system to enter and / or exit intra-day. If the system had been trading on the close, consider pre-computing the price at which an intra-day signal would be generated, then use a limit order for execution. Provided the action price is unique and is a function only of trade price, almost without regard for the complexity of the calculation, the action price can be pre-computed using root finding techniques, such as those described in *Quantitative Trading Systems*.
- Redesign the system to use non-standard parameter values. If the system had been using a common lookback length, such as 20 bars in a standard deviation, see if using some other length returns the system to profitability.
- Redesign the system using adaptive parameters. If the system had been using fixed percentage profit targets, try using targets based on recent volatility.
- Redesign the system using non-standard indicators. If the system had been using simple moving averages, try exponential, weighted, or adaptive averages.
- Redesign the system using different transformations. If an indicator had been ordinary position-in-range, such as the stochastic oscillator, consider using a different position-in-range indicator, such as percent rank.
- Redesign the system using more sophisticated ideas. Try consolidating indicators using principal component analysis.
- Look for trading efficiencies. If the system had been using stop orders and experiencing adverse slippage, see if it can be converted to limit orders where fills will be at your price.
- Reduce frictional costs of slippage and commission.
- Trade other markets. Emerging markets have more pricing inefficiencies than developed markets. But do not greatly sacrifice liquidity or increase counterparty risk.
- Closely monitor system health and adjust position size accordingly. Prepare for an increasing proportion of losing trades and for larger losses by reducing your position size. But do not increase leverage beyond your comfort level in an attempt to increase return.
- Use multiple time frames. Use longer time frames as filters to help distinguish between good and poor entries.

- Use auxiliary data. Use other data series, such as interest rate, volatility, or commodity series, as filters.
- Use diffusion indexes. Let the *wisdom of the crowds* help. Tally the number of advancing and declining issues from among the constituents of an index as an indicator.
- Use different models. If your system has been based on patterns of prices, look into models based on indicators, or seasonality.
- Prepare for the time when trading is no longer profitable.

References

BOOKS

Bandy, Howard, *Introduction to AmiBroker*, Second Edition, Blue Owl Press, 2012
—*Modeling Trading System Performance*, Blue Owl Press, 2011.
—*Quantitative Trading Systems*, Second Edition, Blue Owl Press, 2011.

Connors, Larry, and Cesar Alvarez, *Short Term Trading Strategies that Work*, Trading Markets, 2008.
—*High Probability ETF Trading*, Trading Markets, 2009.
—*How Markets Really Work*, Wiley, 2012.

Derman, Emanuel, *Models.Behaving.Badly*, Free Press, 2012.

Savage, Sam, *The Flaw of Averages, Why We Underestimate Risk in the Face of Uncertainty*, Wiley, 2012.

Silver, Nate, *The Signal and the Noise*, Penguin Press, 2012.

WEBSITES

AmiBroker, *http://www.AmiBroker.com/*

Bandy, Blue Owl Press Blog, *http://www.BlueOwlPress.com/WordPress/*

DTNIQ, *http://www.dtniq.com/index.cfm*

Norgate Premium Data, *http://www.premiumdata.net/*

Program Listings

ABOUT THE PROGRAMS AND LISTINGS

If you wish, you can download the code for each of the program listings.

This book was written over a period of about one year. As drafts of chapters and programs were written, revised, and rewritten, the programs and the accompanying charts and tables went through many versions. It is possible—even likely—that there will be differences between the code in the book and the code you download. The results you get when you run either version on your own computer, with your own data, with your own coffee mug on your own desk, and with your own music may—probably will—be different than those in the book.

I can assure you that the results shown in the book did come from some version of the program shown in the book. If your results are different, it is probably caused by one or more of:

- The data. I used Norgate Premium Data for all end-of-day data.
- The settings. I tried to keep the settings consistent. And, when possible, I tried to control the settings from the afl rather than the settings dialog.
- The parameters. Many of the variables are assigned values in a fairly wide range. The version of the code that was pasted into the book or saved in the download directory might have unusual values.

If there are compiler errors, check the code you are giving AmiBroker for:

- Typographical errors. Those caused by mis-typing.
- Characters AmiBroker does not process correctly. In particular, look at the "double quote marks." AmiBroker wants them to be straight. Some programs, including InDesign used to format the book, translate them into curved marks, with the right and left different.

- Line breaks. AmiBroker ignores all "white space," so statements that have a carriage return in them are interpreted properly. But sometimes the cut-and-paste process breaks a line in a way that confuses the compiler.

In any event, these are intended to be educational, not ready-to-trade, systems. Work through each example. You will be able to come reasonably close to my results. But even more important is understanding what I have written and using it to develop your own systems.

DOWNLOADABLE CODE

Each of the programs in the book that has a listing number can be downloaded. Go to the book's website:

www.MeanReversionTradingSystems/Programs

    Listing 3.1 .... Buy After an N-Day Sequence
    Listing 3.2 .... N-Day Multiposition System
    Listing 4.1 .... Program Template
    Listing 5.1 .... Re-scale a Series
    Listing 5.2 .... Z Transformation
    Listing 5.3 .... Softmax
    Listing 5.4 .... Moving Average Crossover
    Listing 5.5 .... Detrended Price Oscillator
    Listing 5.6 .... Position in Range of DPO
    Listing 5.7 .... Trading System Based on PIR of DPO
    Listing 5.8 .... Generate DV2 Table
    Listing 5.9 .... Future Performance
    Listing 5.10 .... Replicate RSI using Two Series
    Listing 5.11 .... Replicate RSI using EMA
    Listing 5.12 .... Demonstrate EMA's Lambda
    Listing 5.13 .... Allow Fractional Lookback Periods
    Listing 5.14 .... Custom RSI Function
    Listing 5.15 .... Trading System using Custom RSI
    Listing 6.1 .... Test Moving Average Exit
    Listing 6.2 .... Test RSI Indicator Exit
    Listing 6.3 .... Test Z-Score Indicator Exit
    Listing 6.4 .... Test N-Day Holding Period Exit
    Listing 6.5 .... Test First Profitable Open Exit
    Listing 6.6 .... Test Profit Target Exit
    Listing 6.7 .... Test Dynamically Computed Profit
    Listing 6.8 .... Test Trailing Exit

Glossary

Listing 7.1 .... Entry Market on Close
Listing 7.2 .... Enter Next Day Open V1
Listing 7.3 .... Enter Next Day Open V2
Listing 7.4 .... Enter Next Day Open V3
Listing 7.5 .... Enter using Limit Order
Listing 7.6 .... Enter using Stop Order
Listing 7.7 .... Future Leak
Listing 8.1 .... RSI(2) for Filter Testing
Listing 8.2 .... Set a Filter Based on ATR Value
Listing 8.3 .... Test Moving Average as Filter
Listing 9.1 .... 3 Day High Low System—Long Trades
Listing 9.2 .... 3 Day High Low System—Short Trades
Listing 9.3 .... 3 Day High Low System—Aggressive Trades
Listing 9.4 .... Regime Change System
Listing 9.5 .... Connors RSI Pullback System
Listing 9.6 .... Dual Time Frame System—Daily Component
Listing 9.7 .... Dual Time Frame System—Both Components
Listing 9.8 .... Find Surrogates
Listing 9.9 .... Trade SPY from VIX Signals

Index

Adaptive
    Average True Range (ATR), 132
    Moving average, 89, 118-120
    Parameters, 225
Advance-decline, 89, 100-103, 105
Adverse *See* Maximum adverse excursion
Aggressive, 147, 174, 181-191
Alvarez, Cesar, 174, 178, 198
AmiBroker
    Book, 11, 228
    Code *See* Program Listings
    Errors, 231-232
    Explore, 104
    Trading system development platform, 12, 26
    Trademark, 4
Analysis
    Data, 70, 76
    Principle component, 80, 112, 225
    Risk, 147, 167
    Statistical, 198
    Trading system, 23-48, 173
    Validation, 218, 220
    Visual, 113
Anticipating
    Action, 142, 146
    Future data, 34, 81
Artifact
    Chart, 23
    Exponential average length, 107

Average True Range (ATR)
    Filter, 148-159
    Profit target, 132-133
    Trailing exit, 134-135

Backtest
    Hypothetical performance, 5
    Insufficient for validation, 5, 147, 172, 216
    Measure of model fitness, 34
    Settings, 142-143
    Too good, 145
    Trade results, 156
Bandy, Howard, 227
Best estimate set, 36, 156
Bias
    Future performance, 218
    Information, 20
    Judgement, 23
    Prices, 18-19, 56
    Selection, 56, 99
    Survivorship, 19, 99, 200
    Transformation, 76
Black-Scholes, 163-164
Bollinger bands, 33, 72
Bottoms
    Easier to detect, 19
    Finding with ZigZag, 57, 117-139
Bounded indicator *See* Indicator, bounded
Breakout system, 16, 23, 41, 224

Chandelier exit, 133
Cherry picking, 217-218
Clip, 82
Compound Annual Return (CAR)
    Performance metric, 37, 43, 49, 51, 56
    Risk and, 158
Compounding equity, 20-21, 37, 67, 208, 219

Confidence
    Data range, 84
    Drawdown level, 44
    Objective function, 45
    Trading system, 17, 24-26, 31, 37, 40, 43, 65, 112, 173, 218
Connors
    Larry, 174, 178, 198
    RSI, 198-200
Correlated
    Trades, 139, 201, 217-222
    VIX and SPY, 209, 215
Crossover, moving average, 20, 89-91, 146
CSS Analytics, 94
Cumulative Distribution Function (CDF), 73-75
Curse of dimensionality, 45-46, 219

Derman, Emanuel, 31, 227
Detrend
    Price oscillator, 91-94
    Stationarity, 87
Diffusion index, 99-105, 226
Dimensionality, 45-46, 219
Distribution
    Data, 34
    Indicators, 72-76, 103-104
    Moments, 33
    Profit and Risk, 43-44, 47, 158, 163, 169, 209, 218
    Statistical, 72
    Tail, 139
    Trades, 34-37, 158, 165, 219
    Transformation, 80, 85, 88, 95-96, 99, 151, 153
Dominant, 192-196
Drawdown
    Curves *see* Each system
    Intra-trade, 32, 71, 114, 128
    Monte Carlo, 34
    Multi-system, 218-221
    Objective function, 42-43, 51, 113
    Position size, 37, 163
    Risk, 16-17, 20-21, 25-26, 30, 36, 113-115, 147, 216
    Stop trading, 47-48

Dual time frame, 201-209
DV indicator, 94-98, 146
Dynamic
    Indicator, 121
    Position size, 11, 36, 44, 147, 163
    Process, 26, 37
    Profit target, 131-133

Eddington, Arthur, Stanley, 223
Efficient market, 223-224
Entropy, 223
Expectancy, 217, 219
Expectations, 15-16, 30, 34-36, 48, 81, 130, 143, 218
Exponential moving average
    Calculation, 105-112
    Lag, 88
    Use, 89-90, 225

Fibonacci, 160
Filter, 139, 148-163, 173-174, 179-180, 204-209, 225-226
Fisher transformation, 85-86
Future leak, 145

Health, system *See* System health
Historical
    Data, 25, 43, 100
    Indicators, 113
    Patterns, 115
    Volatility, 164
Holding period
    Lag and, 89
    Risk and, 37, 171-172
    Test data, 56
    Trading system, 42, 49, 114, 124-127, 138-139, 173
Hyperbolic tangent, 86-87

Index
    Diffusion, 99-104, 226
    Relative strength *See* Relative Strength Index (RSI)
    Tradable, 15-18, 26, 148, 164, 200

Indicator
    Bounded, 13, 72-73, 79, 198
    Discussed, 69-115
    Entry, 16, 141, 144-145
    Exit, 121-124
    Non-symmetric, 19-20
    Reverse engineered, 146
    Unambiguous, 24
    Unbounded, 13, 72, 79
Inefficient market, 23

Janeczko, Tomasz, 4-5

KISS, 7, 20

Lambda
    Exponential moving average, 107-112
    Softmax, 83-87
Linear
    Regression, 97-87
    Transformation, 77-79, 82-86, 95, 99
Logic
    Combined, 221
    Rules, 24, 28, 34, 36, 118, 120, 124, 138-139, 159, 173, 208
    Synchronization, 30, 41-44, 47-48, 224
Logistic function, 82-84

Maximum adverse excursion (MAE), 36, 100, 102, 135, 220, 225
Maximum favorable excursion (MFE), 33, 100, 102, 128
Mean
    Arithmetic, 33, 80, 84, 89-91, 118, 122-123
    Detrending using, 87, 91-94
    Geometric, 168
    Reversion to, 15-17, 41, 53, 65
Metric *See* Objective function
Model
    Rules plus parameters, 192
    Procedures, 17-19, 23-49
    Signals, 220
    Synchronization, 30, 41-44, 47-48, 223-224

Modeling Trading System Performance, 11-12, 20, 25, 30, 44, 67, 113, 124, 138, 148, 157, 163, 167, 209, 227
Monte Carlo simulation, 11, 30, 34, 37, 44, 147, 218-219

Normal, distribution
    pdf example using, 73-75
    Unimportant for trading systems, 72
Normalization *See* Standardization
    Risk, 156, 158, 163, 167, 169, 209

Objective function, 25-29, 41-44, 45, 49, 91, 173, 202
Optimization, 28, 30, 40, 45, 51, 54, 161, 192
Option
    Expiration, 20
    Tradable, 18, 139, 148, 163-172
Oscillator, 72, 91-94, 225
Overbought, 73, 82
Oversold, 36, 65, 73, 82, 92, 173, 201-209

Parabolic, 133
Percentile, 159, 169
Percent rank, 95-99, 103, 149, 151, 156, 198-203
Position size, 11, 25-26, 30-31, 35-37, 43-44, 47-48, 139, 147-172, 173, 209
Probability, 32, 35, 37, 44, 71, 158-159, 168, 174, 178, 219
Probability density function (pdf), 32, 73-74
Program Listings
    Listing 3.1 Buy After an N-Day Sequence, 50-51
    Listing 3.2 N-Day Multiposition System, 63-64
    Listing 4.1 Program Template, 68
    Listing 5.1 Re-scale a Series, 79
    Listing 5.2 Z Transformation, 80-81
    Listing 5.3 Softmax, 85
    Listing 5.4 Moving Average Crossover, 89-90
    Listing 5.5 Detrended Price Oscillator, 91
    Listing 5.6 Position in Range of DPO, 92-93
    Listing 5.7 Trading System Based on PIR of DPO, 93-94
    Listing 5.8 Generate DV2 Table, 95-96
    Listing 5.9 Future Performance, 100-101
    Listing 5.10 Replicate RSI using Two Series, 106
    Listing 5.11 Replicate RSI using EMA, 106-107

Listing 5.12 Demonstrate EMA's Lambda, 108
Listing 5.13 Allow Fractional Lookback Periods. 108-109
Listing 5.14 Custom RSI Function, 109-110
Listing 5.15 Trading System using Custom RSI, 110-112
Listing 6.1 Test Moving Average Exit, 119-120
Listing 6.2 Test RSI Indicator Exit, 121-122
Listing 6.3 Test Z-Score Indicator Exit, 122-123
Listing 6.4 Test N-Day Holding Period Exit, 125-126
Listing 6.5 Test First Profitable Open Exit, 126-127
Listing 6.6 Test Profit Target Exit, 129
Listing 6.7 Test Dynamically Computed Profit, 132-133
Listing 6.8 Test Trailing Exit, 134-135
Listing 7.1 Entry Market on Close, 142
Listing 7.2 Enter Next Day Open V1, 143
Listing 7.3 Enter Next Day Open V2, 143
Listing 7.4 Enter Next Day Open V3, 144
Listing 7.5 Enter using Limit Order, 144
Listing 7.6 Enter using Stop Order, 145
Listing 7.7 Future Leak, 145
Listing 8.1 RSI(2) for Filter Testing, 149
Listing 8.2 Set a Filter Based on ATR Value, 153
Listing 8.3 Test Moving Average as Filter, 159-161
Listing 9.1 3 Day High Low System—Long Trades, 174-175
Listing 9.2 3 Day High Low System—Short Trades, 178-179
Listing 9.3 3 Day High Low System—Aggressive Trades, 182-187
Listing 9.4 Regime Change System, 193-195
Listing 9.5 Connors RSI Pullback System, 198-200
Listing 9.6 Dual Time Frame System—Daily Component, 201-202
Listing 9.7 Dual Time Frame System—Both Components, 204-206
Listing 9.8 Find Surrogates, 211
Listing 9.9 Trade SPY from VIX Signals, 213

Quantitative Trading Systems, 11, 25, 43, 142, 146, 225, 227

Radtke, Matt, 198
Regime change, 16, 192-198
Regression
    to the Mean, *See* Reversion to the Mean
    Indicator, 41, 97, 166

Relative Strength Index (RSI)
    Calculation, 104-112, 146
    Characteristics, 85-86
    Use, 121-122, 124, 148-149, 153, 159-163, 192-209, 213-215
Re-scale, 75-80
Reversion to the Mean
    Concept, 16
    Techniques, the entire book
Risk

Savage, Sam, 227
Self correcting, 191
Sigmoid, 82-83, 86-87
Silver, Nate, 21, 227
SoftMax, 83-87
Standard deviation, 80, 84, 168
Standardization, 76, 80, 84, 86
System health, 21, 30, 36-37, 41-44, 48, 148, 221, 225

Testing procedure
    Bias, 18
    Date range, 67
    Issues, 56, 67

Z score, 122-123
Z transformation, 80-81